Energy Management:
An Integrated Approach

Energy Management: An Integrated Approach

Maggie Klein

Larsen & Keller
www.larsen-keller.com

Energy Management: An Integrated Approach
Maggie Klein
ISBN: 978-1-64172-149-3 (Hardback)

Larsen & Keller
Published by Larsen and Keller Education,
5 Penn Plaza,
19th Floor,
New York, NY 10001, USA

Cataloging-in-Publication Data

Energy management : an integrated approach / Maggie Klein.
p. cm.
Includes bibliographical references and index.
ISBN 978-1-64172-149-3
1. Power resources--Management. 2. Energy conservation. 3. Energy industries--Management.
4. Energy policy. I. Klein, Maggie.
TJ163.2 .E54 2019
333.79--dc23

Printed and bound in China.

For more information regarding Larsen and Keller Education and its products, please visit the publisher's website www.larsen-keller.com

Table of Contents

	Preface	**VII**
Chapter 1	**Introduction to Energy**	**1**
	a. Renewable Energy	3
	b. Non-renewable Energy	7
Chapter 2	**Energy Production**	**15**
	a. Production of Electrical Energy	15
	b. Production of Renewable Energy	38
Chapter 3	**Energy Supply**	**55**
	a. Energy Policy	59
	b. Energy Consumption	64
	c. Domestic Energy Consumption	71
	d. Electric Power Consumption	82
Chapter 4	**Energy Storage**	**86**
	a. Pumped-storage	96
	b. Superconducting Magnetic Energy Storage	102
	c. Flywheel Energy Storage	110
	d. Battery Energy Storage System	115
	e. Gasholder	116
	f. Thermal Energy Storage	122
	g. Power-to-Gas	126
	h. Fuel Cell	135
Chapter 5	**Energy Management**	**147**
	a. Facility Management	157
	b. Industrial Energy Management	163
	c. Energy Management Challenge	165
Chapter 6	**Energy Conservation**	**167**
	a. Heat Rate	170
	b. Rebound Effect	175
	c. Fuel Saving Device	179
	d. Superinsulation	188
	e. Energy Recycling	190
	f. Radiator Reflector	193
Chapter 7	**Renewable Energy**	**195**
	a. Wind Power	211
	b. Biomass Energy	211
	c. Hydropower	213

d. Geothermal Energy 213
e. Solar Energy 226

Chapter 8 **Energy Economics and Efficiency** **228**
a. Energy Economics 228
b. Energy Efficiency 235
c. Energy Efficiency Gap 238
d. Fuel and Fuel Efficiency 241
e. Energy Efficiency in Transport 244

Permissions

Index

Preface

The planning and operation of energy production and consumption is known as energy management. Its principal objectives are climate protection, resource conservation and cost savings. It is closely associated with production management, logistics and environmental management. The application of energy management to operations involves the areas of facility management, energy procurement, maintenance, energy efficiency management, etc. Some of the potential strategies in this area include passive strategy, strategy of long-term and short-term profit maximization, financially attractive energy measures, etc. Energy management is an upcoming field of science that has undergone rapid development over the past few decades. While understanding the long-term perspectives of the topics, the book makes an effort in highlighting their impact as a modern tool for the growth of the discipline. Coherent flow of topics, student-friendly language and extensive use of examples make this book an invaluable source of knowledge.

To facilitate a deeper understanding of the contents of this book a short introduction of every chapter is written below:

Chapter 1, Energy is the quantity that should be transferred to an object to heat it or perform any other work on it. It can be divided into two categories, renewable energy and non renewable energy. The aim of this chapter is to provide an easy understanding of the varied facets of energy. **Chapter 2**, Energy production is the field that focuses on the production of energy from natural resources. This chapter has been carefully written to provide an easy understanding of the varied facets of energy production such as the production of electrical energy from batteries, production of non renewable energy, production of renewable energy, etc. **Chapter 3**, The delivery of fuels for the purpose of consumption is termed as energy supply. This chapter discusses in detail about energy supply, energy requirement, domestic energy consumption, world energy consumption, etc. in the present day scenario. **Chapter 4**, The process of storing or capturing energy for future use is called energy storage. The topics elaborated in this chapter will help in gaining a better perspective about the different methods of energy storage such as pumped-storage, flywheel energy storage, battery energy storage system, power-to-gas, etc. **Chapter 5**, The management of energy production and consumption via planning and operation is termed as energy management. The chapter closely examines the key concepts of energy management to provide an extensive understanding of the subject. Topics like facility management, industrial energy management, energy management challenge, etc. have been detailed herein. **Chapter 6**, The process of energy consumption achieved through the efficient use and reduction in the consumption of energy is known as energy conservation. All the diverse aspects of energy conservation such as heat rate, rebound effect, fuel saving device, superinsulation, etc. have been carefully analyzed in this chapter. **Chapter 7**, The energy produced from renewable resources which is easily replenished is called renewable energy. Some sources of renewable energy are wind, sunlight, rain, tides, geothermal heat, etc. The topics included in this chapter on wind power, biomass, hydropower, solar energy, etc. discuss the major renewable energies and their generation. **Chapter 8**, The study of the supply and use of energy in a society falls under the discipline of energy economics. In an energy conversion machine, the ratio between the useful generated output and the input is known as energy efficiency. The following chapter elucidates the fundamentals of energy efficiency gap, fuel and fuel efficiency, energy efficiency in transport, etc. as well as the principles of energy economics for a close understanding of energy economics and efficiency.

I would like to share the credit of this book with my editorial team who worked tirelessly on this book. I owe the completion of this book to the never-ending support of my family, who supported me throughout the project.

Maggie Klein

1

Introduction to Energy

Energy is the quantity that should be transferred to an object to heat it or perform any other work on it. It can be divided into two categories, renewable energy and non renewable energy. The aim of this chapter is to provide an easy understanding of the varied facets of energy.

Energy lights our cities, powers our vehicles, and runs machinery in factories. It warms and cools our homes, cooks our food, plays our music, and gives us pictures on television.

Energy is defined as the ability or the capacity to do work.

We use energy to do work and make all movements. When we eat, our bodies transform the food into energy to do work. When we run or walk or do some work, we 'burn' energy in our bodies. Cars, planes, trolleys, boats, and machinery also transform energy into work. Work means moving or lifting something, warming or lighting something. There are many sources of energy that help to run the various machines invented by man.

The discovery of fire by man led to the possibility of burning wood for cooking and heating thereby using energy. For several thousand years human energy demands were met only by renewable energy sources—sun, biomass (wood, leaves, twigs), hydel (water) and wind power.

As early as 4000–3500 BC, the first sailing ships and windmills were developed harnessing wind energy. With the use of hydropower through water mills or irrigation systems, things began to move faster. Fuelwood and dung cakes are even today a major source of energy in rural India. Solar energy is used for drying and heating.

With the advent of the Industrial Revolution, the use of energy in the form of fossil fuels began growing as more and more industries were set up. This occurred in stages, from the exploitation of coal deposits to the exploitation of oil and natural gas fields. It has been only half a century since nuclear power began being used as an energy source. In the past century, it became evident that the consumption of non-renewable sources of energy had caused more environmental damage than any other human activity. Electricity generated from fossil fuels such as coal and crude oil has led to high concentrations of harmful gases in the atmosphere. This has in turn led to problems such as ozone depletion and global warming. Vehicular pollution is also a grave problem.

There has been an enormous increase in the demand for energy since the middle of the last century as a result of industrial development and population growth. World population grew 3.2 times between 1850 and 1970, per capita use of industrial energy increased about twentyfold, and total world use of industrial and traditional energy forms combined increased more than twelvefold.

Due to the problems associated with the use of fossil fuels, alternative sources of energy have become important and relevant in today's world. These sources, such as the sun and wind, can never be exhausted and are therefore called renewable. Also known as non-conventional sources of energy, they cause less emission and are available locally. Their use can significantly reduce chemical, radioactive, and thermal pollution. They are viable sources of clean and limitless energy. Most of the renewable sources of energy are fairly non-polluting and considered clean. However, biomass is a major polluter indoors.

Renewable energy sources include the sun, wind, water, agricultural residue, fuelwood, and animal dung. Fossil fuels are non-renewable sources. Energy generated from the sun is known as solar energy. Hydelis the energy derived from water. Biomass – firewood, animal dung, and biodegradable waste from cities and crop residues – is a source of energy when it is burnt. Geothermal energy is derived from hot dry rocks, magma, hot water springs, natural geysers, etc. Ocean thermal is energy derived from waves and also from tidal waves.

Through the method of co-generation a cleaner and less polluting form of energy is being generated. Fuel cells are also being used as cleaner energy source. In India a number of initiatives have been taken. A good example is the model village of Ralegaon Siddhi.

Measurement of Energy

One of the basic measuring blocks for energy is called a Btu or British thermal unit. Btu is defined as the amount of heat energy it takes to raise the temperature of 1 pound of water by 1 degree Fahrenheit, at sea level. One Btu equals about one black-tip kitchen match. It takes about 2000 Btu to make a pot of coffee.

Energy can also be measured in joules (pronounced the same way as 'jewels'). One joule is the amount of energy needed to lift 1 pound about 9 inches. It takes 1000 joules to equal a Btu. It would take 2 million joules to make a pot of coffee.

Joule is named after an English physicist named James Prescott Joule who lived from 1818 to 1889. He discovered that heat is a type of energy.

Around the world, scientists measure energy in joules rather than Btu. It is much like people around the world using the metric system, meters and kilograms. Like in the metric system, you can have kilojoules: 'kilo' means 1000, therefore, 1000 joules = 1 kilojoule = 1 Btu.

Total commercial energy consumption has been growing tremendously since the last decade. Per capita commercial energy consumption in low-income countries have more than doubled. About 15% of the world's population living in the wealthy industrialized nations consume over half the energy used in the world. The number of motor vehicles in use worldwide has more than doubled since 1970.

In some respects, the global energy system has evolved in a cleaner direction in the last 25 years. The share of world primary energy derived from natural gas – the cleanest fossil fuel has increased by more than 25%. So has the use and generation of renewable energy sources.

Still, the overall efficiency of energy production remains extremely low: on average, more than 90% of energy consumed is lost or wasted in the process of conversion from raw materials such as coal to the final energy service such as the light to read a book. The main problem isn't that we use energy, but how we produce and consume energy resources. What we really need are energy sources that will last forever and can be used without polluting the environment. Conserving energy has become the need of the day be it in the transport, household, or industrial sectors.

Renewable Energy

Renewable energy is a source of energy that can be replenished or replaced within a human's lifetime. The important of renewable energy comes down to sustainability. Since the industrial revolution, global energy demands have skyrocketed and our current dependence on fossil fuels, a non-renewable source of energy, to meet those needs is not something that is sustainable in the long term since we are using up the resources faster than the earth is producing more. It is reasonable to assume that our energy needs will only continue to increase over time as countries population continue to grow, so finding alternatives is important for society now, and especially for future generations.

Sources of Renewable Energy

Some of the emerging sources for alternative energy are in solar, wind, and biomass. Other options include hydropower, tidal, wave and geothermal energy, though they have not had as much growth and mainstream adoption as the others.

Solar Power

- Solar power consists of capturing sunlight in special photovoltaic cells that convert sunlight into electricity. That electricity is then typically stored in batteries for use later.

- Solar power as a source of energy is especially exciting because it produces no pollution during its use and is completely silent. The only pollution created is during the actual manufacture of the cells and is a minuscule amount considering its other environmental benefits.
- Solar power also scales incredibly well. You can use a single small cell to power a calculator, or place an array on a home in the city to supplement or replace reliance on the power grid, up to a large scale solar farm in the dessert to act as a regional power plant.
- It works great in remote locations where running power lines would be difficult or costly. The use of solar panels to power satellites in space is a great example of the ways solar can provide power to places not hooked up to the power grid (and in this case impossible to hook up)

Of course solar is not without its own drawbacks, cost to get it all set up being one of the largest. Cost however continues to come down as the technology matures and tax credits exist in many areas to encourage people to make the initial investment. Solar cells efficiency is also adversely affected by cloudy overcast weather and completely useless during the night. In spite of these negatives though, there is huge potential in the future of solar energy, especially when you consider statistics like the worlds current energy consumption could be taken care for 27 years from the energy of a single day of sunlight hitting the earth. Assuming of course we were able to capture that energy. That is pretty amazing to stop and think about how incredible that is though.

Wind Power

- This is actually a pretty old source of energy for humanity, when you think of how long people have used the power of the wind to sail ships across the world's oceans for trade and exploration, of course sailboats and windmills still exist but these days, wind turbines are used to generate electricity rather than simply crushing grain or to pump water.
- Wind turbines can exist offshore or onshore, though most of the time they are placed in remote or rural areas because of noise created and because the room required for the spinning turbines is not well suited for urban environments.
- Wind power has the same advantages of solar as far as its source being plentiful and completely renewable, with no harmful air or water pollution from its use. It is not as quiet as solar however, and many people complain about the aesthetics of a bunch of turbines being an eye sore or ruining an area's natural beauty. While they might support wind power for the bigger picture of cleaner energy and lessening reliance on our dwindling supply of fossil fuels it becomes a case of a " not in my backyard" sort of mentality.
- Not everyone however shares that opinion and it's worth noting that as of 2014 Denmark generated 40% of its electricity from wind power, with a plan to increase that to 50% by 2020 and 84% in 2035 The Danish government has been a pioneer of using wind power since the 1970's and a large share of turbine manufacturers on the market today are Danish.

Biomass Energy

- Biomass refers to using biological material of living or once living organisms for energy. It could be as simple as directly burning it for a heat source, like a log in a fireplace, or by converting it through various processes into some form of bio-fuel such as ethanol or biodiesel.

- The main advantage of biomass as a fuel is in the use of waste material as a fuel source. Reducing waste is just as important as meeting energy needs, so getting that dual benefit is appealing. Some examples of waste as a bio-fuel would be using discarded vegetable oil to create bio-diesel and capturing methane gasses produced by the breakdown of materials in landfills.
- Corn and Soy are also often used to create ethanol. While it is a renewable source, the biggest problem with biomass as a replacement for our energy needs would be it competing with our food needs. We only have so much land that is suitable for growing our food, much less having to share that limited fertile land to grow corn or soy or whatever for our fuel needs as well. Of course as our population continues to grow, needing more of food and energy, this problem becomes even worse.
- It also requires a good amount of water to grow these crops, a resource that in itself is very important for our survival and in short supply in many regions (the current California drought being a great example).
- The last thing to note about biomass as a fuel or heat source is that while it has lower carbon emissions than fossil fuels, it is still not a totally clean source of energy like wind or solar.

Hydro Power

Hydro power provides a clean and renewable power source as it works from harnessing the flow of water in an area. Rivers can be used, but dams and reservoirs can also be built to create a reliable energy source for the plant.

- Unlike solar and wind, water sources are more stable and reliable. This is turn makes the power to the plant more reliable than these other clean energies too.
- There is no mining and no burning associated with water power. The overall rate of pollution released during the production of hydro power is far less than any other fossil fuel type.
- Hydro power plants may have a higher cost to build, but the lifetime costs of a plant work out to be much more cost effective than fossil fuel dependent energy plants.
- While fossil fuels can take thousands of years to renew themselves, water renews its amount on a regular basis.
- Reservoirs can be used for irrigation, public water and as recreational points in a community. The additional use potential of the reservoir makes this for of power even more cost efficient.
- Creating a reservoir damages the marine ecology of an area that then effects all of the other ecologies around it. Even though the ecology is replace with the one in the reservoir, it is not the ecology native to the area.
- There are certain "wild card" factors that naturally occur that can dramatically lessen a hydro-powers production. Factors such as drought, extreme weather and excess evaporation can all cause water levels to drop.

- While the overall lifespan cost of a plant may be less than fossil fuels, it is much higher in cost to create a hydro power plant and maintain it over its first 10 years.
- Hydro power reservoirs pose an imminent danger to the surrounding human populations. The dam can break, or overflow and cause flash flooding easily.

Geothermal Energy

Geothermal energy is one of the cleanest forms of renewable energy to use. It is done by running systems underground to allow the Earth's natural heating and cooling properties to transfer to the liquid passed through the strata.

- This type of renewable energy scores high on the environmentally friendly scale. There is next to no pollution resulting from its use.
- The source of the energy, the Earth's strata, is always replenishing itself.
- It is thought that if all the areas currently able to use geothermal energy did so it could add 2 terawatts of power to our resources.
- Less reliable powers such as solar and wind can't be used to meet the base load need of a community, but geothermal power is highly reliable and can meet that need.
- You can use geothermal power as easily for a home as you can a major building.
- Half of a geothermal plant is underground so you don't need much space to build one.
- Research programs to improve on current technologies are well funded compared to other energy sources.
- There is no way to build something on land, or go underground, without there being an environmental impact. It is small with geothermal plants, but still there.
- Earthquakes and other seismic activity can damage a geothermal system.
- While less land is used, the cost of constructing a geothermal system is much higher than other types of plants.
- The current technology only allows for certain regions to use geothermal power.
- If a geothermal system is not managed correctly the power resource can run dry.

Tidal Energy

Sometimes referred to as ocean power, tidal energy focuses on harnessing power from the ocean's currents and tides and transforming them into energy for land use.

- Tidal energy is dependent on gravitational fields between the Earth, sun and moon. This forms a constantly renewing cycle of energy.
- This is also one of the greenest and cleanest forms of renewable energy known. It doesn't mean that it is without impact when put through a plant, but that the impact is far less than other energy processes.

- The science of predicting tidal currents is very old and highly developed. This makes it a more reliable source of energy than wind or sun which can break predicted patterns easily.
- It can be hard to get wind and solar to produce when there isn't much going on, but tidal energy is very effective when low speed tides are present due to the high density of water.
- Although this is a very new power it is projected that the power plants themselves will have a longer lifespan and requires less maintenance than other types of energy facilities.
- Yes, tidal power is very green and environmentally friendly, but the exact impact of the plant is not known yet as the technology is too new.
- Right now the actually energy facility for tidal power has to be built near land in strong currents. This occupies real estate that could be used in other ways, and represents a potential threat to the environment and ecology there too.
- This is a very expensive form of energy to implement, mostly because the technology is new.

Non-renewable Energy

Non-renewable energy is that source of energy which cannot be used again and again. This sort of energy is sustainable to exhaustion. These non-renewable energies like coal, oil and gas are getting scarce. Energy sources like oil, coal and gas are known as fossil fuels.

The natural resources like oil, coal and gas are formed into the earth's crust in a period of decades. These energy sources cannot be used again and grown in few days. Nuclear energy is the only form of non-renewable energy form which can be reused. 95% of nuclear can be used again to support electricity power plant. Non-renewable energy sources like coal and oil are getting scarce because the supply cannot meet the demand currently. Natural resources like oil and gas are major fuel component of almost all the industries of the world. The problem with most of the energy sources is that the supply is limited to the few hands. These forces can control the prices of these resources. Oil prices are rising at the will of oil producers in the world like OPERA. The downside of non-renewable energies is that these resources are controlled and owned by few countries and the reservoirs are limited. Moreover these energy sources are blamed for global warming. Our dependence on these sources has given way to extreme competition in the international market for the buying and selling of non-renewable energy sources. This is the reason why major economies of the world are switching to renewable energy sources like sunlight, wind, hydroelectricity and tidal power.

Coal

Coal is a black rock that can be burned to produce energy. Coal is made from fossilized plants. It is ranked depending on how much *"carbonization"* it has gone through: peat is the lowest rank of coal as it has gone through the least amount of carbonization, whereas anthracite is the highest rank of coal.

The history of coal mining goes back thousands of years, back to ancient China and Rome. However, it became crucially important during the Industrial Revolution in the 19th and 20th centuries, when it was primarily used to power steam engines, heat buildings and generate electricity. In many ways, the use of coal as a source of power was what enabled the Industrial Revolution.

Despite the fact that coal has been used for so long, it continues to be relatively abundant and cheap to convert to energy. However, coal extraction and combustion are both environmentally hazardous processes which carry with them a number of health concerns. Coal mining, especially surface mining, can be particularly destructive of landscapes and pollute local resources such as water and soil.

What is more, air pollution from coal-fired power plants releases sulfur dioxide, nitrogen oxides, particulate matter and heavy metals. These emissions create a number of dangerous environmental problems such as smog and acid rain, as well as numerous health risks for us, ranging from the diseases of the respiratory tract, cardiovascular, and cerebrovascular system . In addition, coal is also an emitter of CO_2 which contributes to climate change.

While some countries such as the UK have set a date of phasing out coal-fired plants, coal remains a key source of energy around the world.

Globally in 2013 we used 7,876 metric tonnes of coal, this was a 70% increase from 2000.

For example, in the United States, 39% of the country's electricity was produced with coal. To put this in perspective, your average coal train is 2.4 kilometres long and yet it carries barely a day's fuel for a large power plant.

Burning coal

Oil

Another form of non-renewable energy is oil, a very versatile liquid fossil fuel that can be used for energy generation as well as a wide range of other applications. It is usually stored deep beneath the earth's surface. Oil is extracted by drilling into where it is trapped between 2 layers of rock, a pipe is then inserted in the hole to siphon it to the surface.

When the so-called drill rigs are set up, either onshore or offshore, oil can be extracted 24/7 for several decades. Once the oil has been drilled, it must be refined to separate it from several other chemicals.

When it is refined around 45 gallons of petroleum products are created through refinery processing gain. The products created are gasoline, ultra-low sulfur distillate, jet fuel, hydrocarbon gas liquids, heavy fuel oil and heating oil.

The advantages of oil are that it is the fuel that has made modern transportation possible, it is relatively cheap to produce and use. About half of the world's petroleum is converted into gasoline used to fuel our cars. The rest can be processed and used in liquid products such as nail polish and rubbing alcohol, or solid products such as water pipes, shoes, crayons, roofing, vitamin capsules, and thousands of other items. Combined with other chemicals oil provides the basis of both rigid and flexible plastic as well as a number of solvents.

The world used 35 billion barrels of oil in 2016.

While oil's versatility is impressive, there are environmental risks that come with extracting oil and using it. One very important consideration is the risk of spills as oil can destroy local habitats and kill animals. Some of the most renowned environmental disasters are linked to oil spills – for example, the Exxon Valdez and BP Deepwater Horizon oil spill.

Oil's environmental and health impacts are similar to those of coal, particularly when it comes to air pollution and climate change.

Spilled oil

Natural Gas

Natural gas is the second most abundant non-renewable energy source. While coal is a rock and oil is a liquid, natural gas is – as the name suggests – a gas. Natural gas is predominantly methane. It was formed like the other fossil fuels with pressure and heat pressing on organic material over millions of years. Unlike oil and coal deposits, natural gas is not found in big open pockets but rather trapped in rock formations.

The process of extraction, called hydraulic fracturing, is slightly more complicated. It involves high-pressure water being piped underground to break the rock and release the gas. If the rock is too hard, chemicals are used to dissolve the rocks.

When natural gas is initially extracted from the ground, we get "wet natural gas" because it usually contains liquid hydrocarbons and non-hydrocarbon gases. To separate methane and other useful gases from the wet natural gas, it has to be processed.

The total amount of natural gas consumed in the world in 2010 was 113 trillion cubic feet.

The biggest advantage of natural gas is the affordability of this source of energy – that's what allows its widespread use for heating and cooking.

Compared to coal and oil, natural gas emits a lot less CO_2. From this point of view, it is better for our environment. The other substance released is water vapor, which is nowhere near the hazardous air pollutants emitted by the oil and coal industry.

At the same time, natural gas is primarily made up of methane, another greenhouse gas, which is far more potent than CO_2. This means that using and transporting natural gas could potentially release large amounts of methane into the atmosphere, which can have disastrous implications for the planet's climate.

While natural gas is considered the cleanest of fossil fuels, its extraction is problematic. Hydraulic fracturing can create mini-earthquakes and the use of chemicals can contaminate nearby soil and water resources. Mini-earthquakes can be particularly damaging to local communities particularly in terms of the value of their houses and property.

Burning natural gas for cooking

Nuclear Power

Nuclear power is not a fossil fuel but it is a non-renewable form of energy. Nuclear power is considered non-renewable because the material used in nuclear power plants – uranium – is a non-renewable resource.

Nuclear energy is generated from the energy released when atoms of uranium are split during the process of nuclear fission in a nuclear reactor. The type of uranium used for nuclear energy is U-235. This uranium isotope is used because its atoms are easily split apart.

Uranium is a common metal found in rocks worldwide, but nuclear energy needs uranium 235 for nuclear fission. Uranium is extracted from rock and then enriched to become uranium 235 isotope before being made into pellets that are loaded into assemblies of nuclear fuel rods.

The great advantage of nuclear energy is that nuclear power plants do not pollute the air or emit greenhouse gases. Small amount of radioactive material produces a lot of energy, the raw materials are relatively cheap and last a long time.

Having said that, not only are U-235 reserves limited but extracting uranium, as well as disposing of the waste generated from the production of nuclear energy can be damaging to the environment.

The radioactive material is dangerous and any accidents can be particularly costly to both the environment and our health.

Working towers of a nuclear power plant

Alternative to Non-Renewable Energy Sources

There are many active groups that are finding other alternative sources of energy. For instance, there has been increased exploitation and tapping of renewable energy sources. Worldwide there are several renewable energy projects that are undergoing to harness this energy as it is clean, replenished, sustainable and widely available.

Most people argue that even these projects consume fossil fuel in some indirect way to produce electricity. Manufacturing of turbines, generators, blades, solar panels, transformers and several other small components are done in factories which use fossil fuels to produce them. But, renewable energy experts say that energy produced by these plants is much higher than the energy consumed during making of these components.

Although, these energy sources have been in use since last few decades but it is only in the last few years that renewable energy has speed up the pace. Huge investments have been going in these projects and several companies across the world are setting up wind farms, solar farms and hydropower plants across several countries. Nuclear energy has also gain momentum.

Though it is not considered as renewable source due to the fact that it releases some toxic substances in the form of radioactive waste but it is sustainable form of energy. Even after several years, share of renewable energy in world power production is less than 20%. It is however believed that by 2020, renewable energy will have a major share in world's electricity production.

Coal mining as predominant as it is, should be limited to save the environment from many other associated effects aside from pollution. This was the major energy source in the early 19th century before plans were made to tap into renewable energy sources. Oil exploration was once the best trade and investment to make.

Nowadays, any discovery is just an environmental disaster in waiting. Geothermal power is another one that gives out lots of energy. The one bad thing is it is a non-renewable source of energy that has vast environmental effects. When carbon monoxide is released in the air, it bonds with oxygen resulting in reduction of percentage of oxygen in that specific area. This is a gas that has the capability to suffocate as it deprives oxygen from its surrounding to form carbon dioxide.

Future of Non-Renewable Energy Sources

Non-renewable energy sources have fueled the world's industrial complex for far too long. It has reached a point where the word is facing rapid starvation in this sector. There are also other associated effects too which also need to be carefully looked at just to make sure things are running as intended. However, this is not the case. With increased exploitation of these fossil fuels, there are many associated environmental effects like land pollution and air pollution which in turn affect both animal and plant life. The far-reaching consequences of non-renewable sources are inexplicable and the trend has to be reversed soon before it is too late to do anything.

Carbon is a major source of fuel in non-renewable energy sources. When combustion takes place, carbon is mixed with oxygen and form carbon dioxide. It pollutes the environment and is responsible for global warming. In last few years, the concentration of carbon dioxide has only increased in the atmosphere. Not to mention, climate change, acid rain and change in seasons are some other effects that has been observed by many people. With so many problems, scarce resources and rising prices, these resources cannot be used for lifetime. The need of the hour is to look for some alternative sources of energy and protect our environment from such harmful gases.

Advantages of Renewable Energy

1. Renewable Energy is Eco-friendly

It is a clean source of energy, meaning, it has low or zero carbon and greenhouse emission. Fossil fuels emit high levels of greenhouse gas and carbon dioxide, which are greatly responsible for global warming, climate change, and degradation of air quality. Fossil fuels also contribute to sulfur emission to the atmosphere leading to acid rains. Acid rains can cause damage to buildings. Solar and wind power are considered eco-friendly because they emit zero toxic gases to the environment. The use of renewable energy dramatically reduces the dependence on fossil fuel as a source of energy, hence, cutting back on air pollution.

2. It's a Renewable Resource

This implies that they do not deplete over a lifetime and there is zero possibility that they will run out (sustainable source of energy). Sources of energy like fossil fuels (oil, gas, and coal) are considered limited resources and there is strong possibility that they will run out in the future. Renewable energy can help developing countries from over-reliance on fossil fuels. Powerful winds, heat emanating from beneath the earth, sunshine and moving water can guarantee a huge and steady energy supply to a nation for many years.

3. Renewable Energy is a Reliable Source of Energy

In the previous few decades, the use of fossil fuel has sharply increased. This over-reliance on fossil fuels has led to our security being threatened. Fossils fuels are prone to trade disputes, political instabilities, spike in energy prices and unnecessary wars. These variables affect a lot more than a nation's energy policies; they can significantly drain a county's economy.

Although most argue that solar and wind energy are unreliable, a solid infrastructure puts this argument to rest. If solar and wind plants are distributed over a large geographical location, there can be minimal electricity generation interruption because weather disruptions in one location cannot be the same in other locations.

4. Leads to Job Creation

Renewable energy makes real economic sense because it is a cheaper alternative to most traditional sources of energy. Since the inception of renewable energy, new and stable jobs have been added to most world economies. For, instance, in Germany and UK, many jobs have already been created thanks to their relentless efforts to develop and encourage the use of renewable forms of energy. Experts project that with the ongoing rigorous campaigns to embrace renewable energy, thousands of stable jobs will be created.

5. Renewable Energy has Stabilized Global Energy Prices

Change up to renewable sources of energy means stability of energy prices across the globe. This is because the cost of renewable energy depends on the initial cost of installation of renewable energy technologies as opposed to fossil fuels, which increase and decrease depending on the current inflation and availability of the resource. Respective governments would only need to cater to the initial costs and that's it.

6. Less Maintenance of Facilities

Once infrastructure for the harnessing of the renewable resource is laid down, there is low to zero maintenance required. This means that the owners of the facilities will reap big profits while providing cheap electricity to the population.

7. Boosts Public Health

This is a no-brainer. If governments took upon themselves to build more renewable energy facilities, the population would enjoy the health benefits. According to a study by the U.S. Environmental Protection Agency, Americans spend approximately $361.7 to $886.5 billion every year on overall health of the population. A big chunk of this budget goes to mitigate and cure diseases related to fossils fuel use like heart diseases, cancer, and neurological disorders.

Greenhouse, carbon and sulfur compounds emitted by fossils fuels are risky to our health if inhaled over time. This is reason enough to consider renewable energy moving forward.

8. Empowering of People in The Countryside

Renewable energy generation mainly takes place in remote settings. This means that local towns would get a fair share of power generated, ultimately, catalyzing the regeneration of those depressed areas both socially and economically. Electrification of those areas will open up untapped opportunities for development through the advancement of greenhouses using geothermal power, district heating of towns and communities through hot water generated by the energy exploitation of forestry and agricultural biomass.

Disadvantages of Renewable Energy

1. The Electricity Generation Capacity is Still Not Large Enough

There are still challenges to generation of large quantities of power in renewable energy technology compared to traditional forms of energy generation like fossil fuel. Fossil fuel still produces large quantities of electricity today, by far. This, essentially, means that it can't be solely relied upon to power the whole nation.

This means that either we need to set up more such facilities to match up with the growing demand or look out for ways to reduce our energy consumption. This phenomenon indicates that a balance of different energy sources will still prevail for some years to come.

2. Renewable Energy Can be unreliable

Renewable energy technologies totally depend on the weather (for e.g.: sun and wind) to be able to harness any energy. In case atmospheric conditions are not good enough, renewable energy technologies would lack the ability to generate any electricity. This might instigate campaigns by the authorities to reduce energy usage in order to serve the population for a longer period.

3.Low-efficiency Levels

Renewable energy technologies are still significantly new to the market, meaning, they still lack the much-needed efficiency. This poses forecast problems and investors may shy away from investing their money for fear of not getting returns pretty quick.

4. Requires a Huge upfront capital outlay

Setting up renewable energy generation facilities requires a huge financial outlay. Installation of wind turbine, solar panels, and hydroelectricity plants are relatively expensive. These plants require upfront investments to build, have high maintenance expenses and require careful planning and implementation. Also, the electricity generated needs to be delivered to towns and cities, which means additional cost of installing power lines.

References

- What-are-the-4-main-types-of-non-renewable-energy: greentumble.com, Retrieved 30 June 2018
- What-is-renewable-energy: conserve-energy-future.com, Retrieved 19 May 2018
- Advantages-and-disadvantages-of-renewable-energy: conserve-energy-future.com, Retrieved 12 March 2018
- Non-renewable-energy-sources: conserve-energy-future.com, Retrieved 10 April 2018

2

Energy Production

Energy production is the field that focuses on the production of energy from natural resources. This chapter has been carefully written to provide an easy understanding of the varied facets of energy production such as the production of electrical energy from batteries, production of non renewable energy, production of renewable energy, etc.

Production of Electrical Energy

Electrical energy is produced from many different energy sources. Some of these energy sources are renewable and others are non-renewable. The first ever public electric power distribution system appeared in the late 19th century and mostly supplied direct currents, with multiple output voltages, or under the form of alternating current, which reverses the polarity of the current every 20 milliseconds. However, the energy demand and the advantages that derive from the use of standard voltage outputs quickly impose the alternating current as the preferred type for production, transport and distribution.

Most of the electric energy production process takes place in power plants burning fossil fuel, such as coal, natural gas or crude oil, which accounts for the massive amounts of greenhouse gases emitted daily in the Earth's atmosphere. Heat produced during fuel burning is used to evaporate large amounts of water and create Steam that powers a steam turbine, turning the energy extracted during burning fuel into mechanical work. By connecting a mechanical generator to the turbine's drive shaft, the mechanical work can be turned efficiently into electric energy.

However, the electric energy cannot be directly supplied to the consumer in this form, thus it must undertake certain transformations in order to meet the required standard, which state that the electric current must have parameters ranging in certain imposed values. For example, in Europe, most of the countries use alternating electric current that has a nominal voltage of 220 volts and an oscillating frequency of 50 hertz, while the United States prefer an electric current having 110 volts, and a frequency of 60 hertz.

The frequency of the alternating current is usually determined by the construction of the generator, however voltage of the current represents a key aspect in the transport and distribution of the electric energy, thus it is uneconomical and unpractical to build generators that produce different voltages.

Instead, transformers are used to vary the voltage of the electric current in the required domain. Transformers consist basically of two electric circuits formed of two coils, separate electrically by an inductive coupling used to transfer energy between the two. There are two primary types of electric transformers: high-voltage transformers and low-voltage transformers. High-voltage transformers are used to increase the voltage output, while the low-voltage ones are used to decrease it.

The electric energy transport is usually greatly affected by the loss of power in long conductors, as a result of the presence of electrical resistance. Even the best electrical conductors having low electrical resistance can produce enormous power loss over relative small distances. To solve this problem, the electrical current's voltage is being increased with the help of high-voltage transformers, meaning that for a voltage increase of a 10 factor, the current intensity will drop by the same amount, thus reducing the power loss by over one hundred times.

Most of the outlets present in our houses have a mono-phase current configuration, consisting of two electrodes, one used as ground and the other to supply the alternating electric current. However, the transport systems mostly imply the use of three-phase currents that consist of four metal conductors, three for alternating electric current supply, while most of the time the fourth conductor is used as ground, though this is not necessarily required.

The electric currents through the three-phase metal system are out of phase by 120 solid degrees and voltages between the lines usually measure an average of 220 volts.

The use of fossil fuel for electric energy production has determined massive emissions of greenhouse gases; however, in the last decades, attention shifted towards alternative energy sources, such as wind power, solar power and, last but not the least, nuclear power that is quickly becoming the base of electric power production.

There are various methods of electricity generation dependent on types of energy. Among resource energies, coal and natural gas are used to generate electricity by combustion (thermal power), Uranium by nuclear fission (nuclear power), to utilize their heat for boiling water and rotating steam turbine. Among renewable energies, sunlight is directly converted into electricity (photovoltaics), rotation energy by wind is converted into electricity (wind power), rotating water wheel by running water to generate (hydro). Magmatic heat boils underground water to rotate steam turbine to generate (geothermal).

Continuous technology development for them are proceeding to convert resource energies or renewable energies into electricity with less loss. It is also important for the operation of power plant to do maintenance or training of operators.

Thermal Power

Pulverized Coal-fired Power Generation

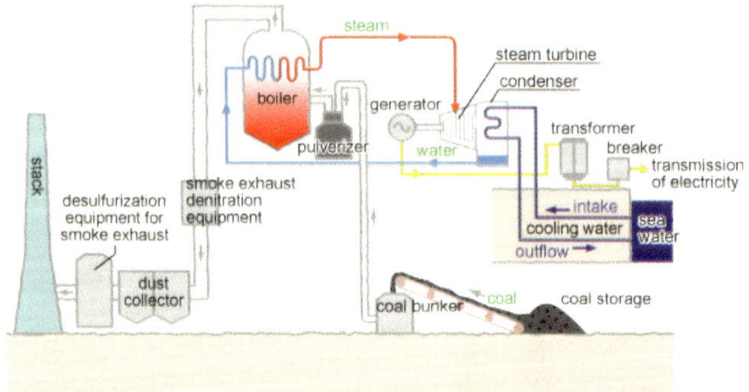

Pulverized coal-fired power generation is currently the major method of coal-fired power generation. Coal is pulverized to fine powder and is burned in the boiler. Heat in the boiler boils the water into steam. The steam pressure turns the steam turbine and the generator generates electricity.

Combined Cycle Power Generation

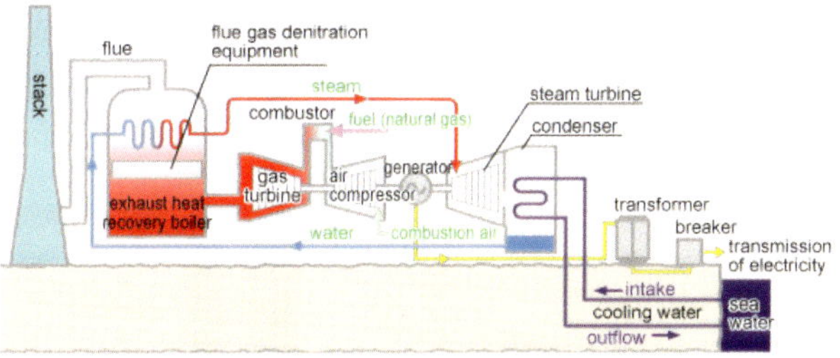

A combined cycle power first generates gas by burning fuel in the compressed air.Pressure of the gas rotates the gas turbine and the generator creates electricity. Moreover, exhaust heat from gas turbine is utilized for boiling water to generate steam, which rotates turbine to generate.

Integrated coal Gasification Combined Cycle (IGCC)

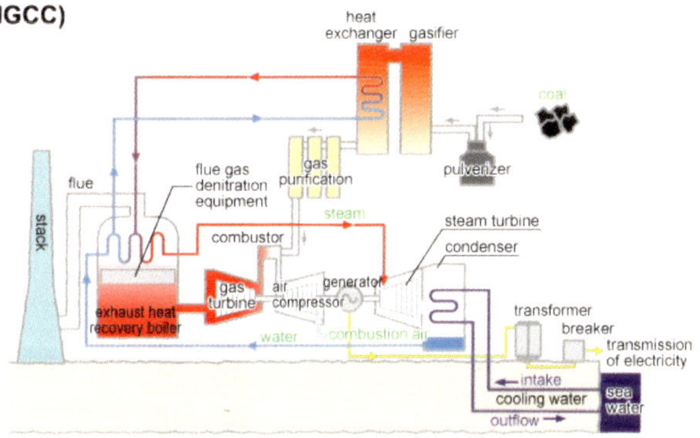

Integrated coal gasification combined cycle (IGCC) gasifies the fuel coal in the gasifier. Gasified fuel is burnt in the compressed air to generate gas. Pressure of the gas rotates the gas turbine to generate electricity. Furthermore the exhaust gas heat from gas turbine is utilized to boil water into steam to generate electricity.

Nuclear Power

Boiling water reactor (BWR)

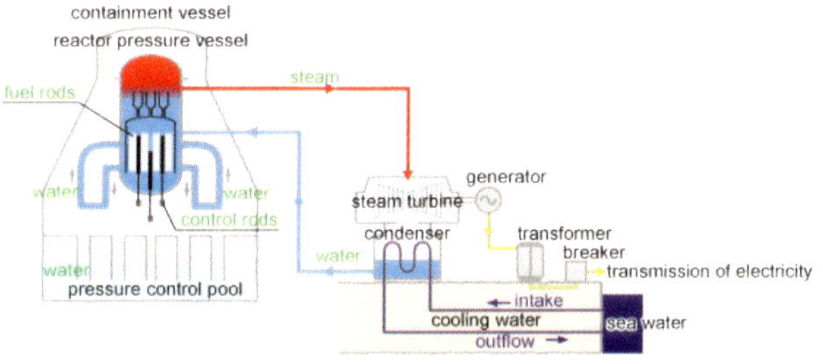

Lightwater means normal water, opposite to heavy water. Heat is produced by nuclear fission in the reactor core, then causes the water to boil, producing steam. The steam is used to rotate the turbine to generate electricity, then is cooled in the condenser by sea water and converted back to liquid water. This water is then returned to the reactor core.

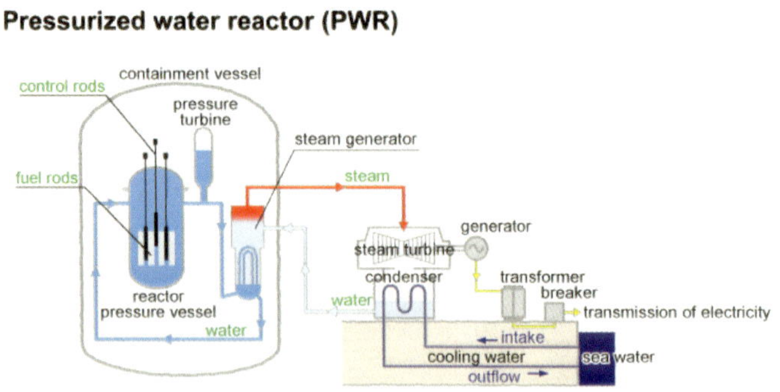

Lightwater means normal water, opposite to heavy water. Heat is produced by nuclear fission in the reactor core but heated water is suppressed before boiling by applying a high pressure. This water with high temperature and pressure is sent to the steam generator, boils the water into steam, and then rotates the turbine to generate electricity in the generator, after that it is cooled in the condenser by sea water and converted back to liquid water. This water is then returned to the steam turbine.

Impact of Generation of Electrical Energy

Production of electricity from any form of primary energy has some environmental effect, and some risk. A balanced assessment of nuclear power requires comparison of its environmental effects with those of the principal alternative, coal-fired electricity generation, as well as with other options. That comparison needs to recognize that waste and decommissioning costs are internalized in nuclear power economics much more fully than, for example, with coal-fired power generation.

Environmental Effects of Electricity Generation

These include the effects of obtaining the fuels from mines, using the fuels, and dealing with the wastes following use of the fuel.

At a uranium mine, ordinary operating procedures normally ensure that there is no significant water or air pollution. The environmental effect of coal mining today is also small except that more extensive areas are often disturbed and may require subsequent rehabilitation, and in certain situations of geology and climate, acid mine drainage due to oxidation of sulfur can be a problem.

Burning any fossil fuel gives rise to carbon dioxide.

Small amounts of radioactivity are released to the atmosphere from both coal-fired and nuclear power stations. In the case of coal combustion small quantities of uranium, radium and thorium present in the coal cause the ash to be radioactive, the level varying considerably. Nuclear power stations and reprocessing plants release small quantities of radioactive gases (e.g. krypton-85 and xenon-133) and iodine-131, which may be detectable in the environment with sophisticated monitoring and analytical equipment but are never at harmful levels. Steps are being taken to reduce further emissions of both fly ash from coal-fired power stations and radionuclides from nuclear power stations and other plants. At present neither constitutes a significant environmental problem.

Disposal and Dispersal of Wastes from Power Generation

The solid high-level waste from nuclear power stations is hot and very radioactive, so must be isolated from people and the environment indefinitely. It is stored for 40-50 years while the radioactivity decays to less than one percent of its original level. Then it will be finally disposed of deep underground and well away from the biosphere. In more than 50 decades of civil nuclear power experience nuclear wastes have not caused any serious health or environmental problems nor posed any real risks to people. There has been no pollution or plausible hazard from such material routinely removed from power stations and nor is any likely, either short- or very long-term.

Intermediate-level waste (radioactive but not requiring cooling) is placed in underground repositories, not necessarily deep, with little delay. Low-level waste is generally buried more conventionally. Radioactive fly ash from coal-fired power stations has in the past had a much greater environmental impact largely because it was not perceived as a problem and appropriate action was measures were not taken. Today most fly ash is removed from stack gases and mixed with bottom ash before being buried where seepage and run-off can be controlled, or the ash may be used with cement in concrete.

Nuclear wastes are certainly a significant part of the nuclear power picture, and need to be managed and disposed properly.

Alternatives for power generation are not without challenges and for a variety of reasons they particularly those from coal combustion have not always been well controlled. Fly ash particularly, but also bottom ash, are often loaded with heavy metals (including uranium and thorium). Fly ash is mostly retained for land disposal today, and bottom ash is normally buried also, but not always securely and without effects on groundwater. Groundwater pollution with arsenic, boron, cobalt and mercury is not unusual, and the US EPA in 2011 listed 181 US coal ash ponds which posed a significant hazard, 47 of these a high hazard and threat to life.

Waste heat produced due to the intrinsic inefficiency of energy conversion, and hence as a by product of power generation, is much the same whether coal or uranium is the primary fuel. The thermal efficiency of coal-fired power stations ranges up to a possible 40 percent, with newer ones typically giving better than 35 percent. That of nuclear stations mostly ranges from 29-38 percent with the common light water reactor today giving about 34 percent.

There is no reason for preferring one fuel over the other on account of the amount of waste heat and consequent water requirements for cooling. This is the case whether power station cooling is by water from a stream or estuary, or using atmospheric cooling towers which evaporate water. However, it is noteworthy that whereas coal-fired power plants tend to be located near a source of coal, nuclear plants can be sited according to cooling requirements, and can more readily make use of lake or sea water for direct cooling. Hence they are less likely to require expensive cooling towers or to deplete supplies of fresh water for evaporative cooling.

For a given level of thermal efficiency and size of plant the cooling requirements for a nuclear plant are slightly greater as it does not lose heat up the stack with combustion products.

In any case the dumped heat need not always be 'waste'. In colder climates district heating and agricultural uses are increasingly found. In France the waste heat from a nuclear plant is used for a crocodile farm. Any such use of waste heat decreases the extent to which local fogs result from its release to the environment in winter. In dryer climates, the rejected heat can be used for desalination to provide potable water.

The main environmental matter relevant to power generation is the production of carbon dioxide (CO_2) and sulfur dioxide (SO_2) as a result of coal-fired electricity generation. When coal of say 2.5 percent sulfur is used to produce the electricity for one person in an industrialized country for one year, then about 9 tonnes of CO_2 and 120 kg of SO_2 are produced.

Sulfur dioxide emissions arise from the combustion of fossil fuels containing sulfur, as many of them do. Released in large quantities to the atmosphere it can cause (sulfuric) "acid rains" in areas downwind.

It is possible to remove a lot of the SO_2 from coal stack gases using flue gas desulphurization equipment, but the cost is considerable.

Oxides of nitrogen (NOx) from fossil fuel power stations operating at high temperatures are also an environmental problem, regardless of the fuel source. If high levels of hydrocarbons are present in the air, nitrogen oxides react with these in sunlight to form photochemical smog. Moreover, oxides of nitrogen have an adverse effect on the Earth's ozone layer, increasing the amount of ultra-violet light reaching the Earth's surface.

Health Effects of Power Generation

Traditionally occupational health risks have been measured in terms of immediate accident, especially fatality, rates. However, today, and particularly in relation to nuclear power, there is an increased emphasis on less obvious or delayed effects of exposure to cancer-inducing substances and radiation.

There are a variety of chemical and mechanical devices that are called batteries, although they operate on different physical principles. A battery for the purposes of this explanation will be a device that can store energy in a chemical form and convert that stored chemical energy into electrical energy when needed. These are the most common batteries, the ones with the familiar cylindrical shape. There are no batteries that actually store electrical energy; all batteries store energy in some other form. Even within this restrictive definition, there are many possible chemical combinations that can store electrical energy.

There are two fundamental types of chemical storage batteries: the rechargeable, or secondary cell, and the non-rechargeable, or primary cell. In terms of storing energy or discharging electricity, they are similar, it is simply a question of whether or not the chemical processes involved permit multiple charging and discharging.

Before answering this question it is also necessary to distinguish between a galvanic cell and a battery. The former is the fundamental unit of electrochemical storage and discharge. A battery is comprised of at least one but possibly many such cells appropriately connected. Because the cell is where the actual action of storage and discharge takes place, this answer will concentrate on what happens at that level.

All electrochemical cells consist of two electrodes separated by some distance. The space between the electrodes is filled with an electrolyte-an ionic liquid that conducts electricity. One electrode--the anode-permits electrons to flow out of it. The other the cathode receives them. The energy is stored in the particular compounds that make up the anode, cathode and the electrolyte-for example, zinc, copper, and SO_4, respectively.

Assuming the battery has acquired its charged condition either by recharging or manufacturing, the aggregate effect of the chemical reactions taking place between the anode and the cathode discharges electricity. The anode undergoes what is known as an oxidation reaction: during discharge two or more ions from the electrolyte combine with the anode to form a compound and release one or more electrons. Simultaneously, the cathode undergoes a reduction reaction wherein the material the cathode is made of, ions, and free electrons combine to form compounds.

Simply put, the chemical reaction at the anode releases electrons and the reaction at the cathode absorbs them. When the electrical path provided by the electrolyte and an external electrical circuit connects the anode and cathode, the two simultaneous reactions proceed and the electrons freed at the anode travel through the external electrical connection and react chemically at the cathode to make the cell function. The cell can continue to discharge until either or both of the electrodes run out of reagents for their respective reactions. In a primary cell this means the end of its useful life, but in a secondary cell it just means it is time for a recharge. For secondary cells the recharge process is the reverse of the discharge process. An external source of direct electrical current supplies electrons to the anode and removes them from the cathode, forcing the chemical reactions into reverse until the cell is recharged.

The above constitutes a simplified explanation of how the electrochemical energy stored in a cell is removed as electrical energy in the process of discharging and restored in the process of recharging a secondary cell. There are many more electrochemical and thermal processes taking place at the same time and for most practical cell combinations packaged in the form of batteries it is not possible to completely characterize all of the processes.

Non-renewable energy comes from sources that will run out or will not be replenished in our lifetimes—or even in many, many lifetimes.

Most non-renewable energy sources are fossil fuels: coal, petroleum, and natural gas. Carbon is the main element in fossil fuels. For this reason, the time period that fossil fuels formed (about 360-300 million years ago) is called the Carboniferous Period.

All fossil fuels formed in a similar way. Hundreds of millions of years ago, even before the dinosaurs, Earth had a different landscape. It was covered with wide, shallow seas and swampy forests.

Plants, algae, and plankton grew in these ancient wetlands. They absorbed sunlight and created energy through photosynthesis. When they died, the organisms drifted to the bottom of the sea or lake. There was energy stored in the plants and animals when they died.

Over time, the dead plants were crushed under the seabed. Rocks and other sediment piled on top of them, creating high heat and pressure underground. In this environment, the plant and animal remains eventually turned into fossil fuels (coal, natural gas, and petroleum). Today, there are huge underground pockets (called reservoirs) of these non-renewable sources of energy all over the world.

Advantages and Disadvantages

Fossil fuels are a valuable source of energy. They are relatively inexpensive to extract. They can also be stored, piped, or shipped anywhere in the world.

However, burning fossil fuels is harmful for the environment. When coal and oil are burned, they release particles that can pollute the air, water, and land. Some of these particles are caught and set aside, but many of them are released into the air.

Burning fossil fuels also upsets Earth's "carbon budget," which balances the carbon in the ocean, earth, and air. When fossil fuels are combusted (heated), they release carbon dioxide into the atmosphere. Carbon dioxide is a gas that keeps heat in Earth's atmosphere, a process called the "greenhouse effect." The greenhouse effect is necessary to life on Earth, but relies on a balanced carbon budget.

The carbon in fossil fuels has been sequestered, or stored, underground for millions of years. By removing this sequestered carbon from the earth and releasing it into the atmosphere, Earth's carbon budget is out of balance. This contributes to temperatures rising faster than organisms can adapt.

Coal

Coal is a black or brownish rock. We burn coal to create energy. Coal is ranked depending on how much "carbonization" it has gone through. Carbonization is the process that ancient organisms undergo to become coal. About 3 meters (10 feet) of solid vegetation crushed together into .3 meter (1 foot) of coal.

Peat is the lowest rank of coal. It has gone through the least amount of carbonization. It is an important fuel in areas of the world including Scotland, Ireland, and Finland.

Anthracite is the highest rank of coal. Anthracite forms in regions of the world where there have been giant movements of the earth, such as the formation of mountain ranges. The Appalachian Mountains, in the eastern part of the United States, are rich in anthracite.

We mine coal out of the ground so we can burn it for energy. There are two ways that we can mine coal: underground mining and surface mining.

Underground mining is used when the coal is located below the surface of the Earth, sometimes 300 meters (1,000 feet) deep—that's deeper than most of the Great Lakes! Miners take an elevator down a mineshaft. They operate heavy machinery that cuts the coal out of the Earth and brings it above ground. This can be dangerous work because cutting coal can release dangerous gases. The gases can cause explosions or make it hard for miners to breathe.

Surface mining is used when the coal is located very near the surface of the earth. To get to the coal, companies must first clear the area. They take away the trees and soil. The coal can then be cut out of the ground more easily. Entire habitats are destroyed during this process.

About half the electricity in the United States comes from coal. It gives power to our lights, refrigerators, dishwashers, and most other things we plug in. When coal is burned, it leaves "byproducts" that are also valuable. We use the byproducts to make cement, plastics, roads, and many other things.

Advantages and Disadvantages

Coal is a reliable source of energy. We can rely on it day and night, summer and winter, sunshine or rain, to provide fuel and electricity.

Using coal is also harmful. Mining is one of the most dangerous jobs in the world. Coal miners are exposed to toxic dust and face the dangers of cave-ins and explosions at work.

When coal is burned, it releases many toxic gases and pollutants into the atmosphere. Mining for coal can also cause the ground to cave in and create underground fires that burn for decades at a time.

Petroleum

Petroleum is a liquid fossil fuel. It is also called oil or crude oil.

Petroleum is trapped by underground rock formations. In some places, oil bubbles right out of the ground. At the LaBrea Tar Pits, in Los Angeles, California, big pools of thick oil bubble up through the ground. Remains of animals that got trapped there thousands of years ago are still preserved in the tar!

Most of the world's oil is still deep under the ground. We drill through the earth to access the oil. Some deposits are on land, and others are under the ocean floor.

Once oil companies begin drilling with a "drill rig," they can extract petroleum 24 hours a day, seven days a week, 365 days a year. Many successful oil sites produce oil for about 30 years. Sometimes they can produce oil for much longer.

When oil is under the ocean floor, companies drill offshore. They must build an oil platform. Oil platforms are some of the biggest manmade structures in the world!

Once the oil has been drilled, it must be refined. Oil contains many chemicals besides carbon, and refining the oil takes some of these chemicals out.

We use oil for many things. About half of the world's petroleum is converted into gasoline. The rest can be processed and used in liquid products such as nail polish and rubbing alcohol, or solid products such as water pipes, shoes, crayons, roofing, vitamin capsules, and thousands of other items.

Advantages and Disadvantages

There are advantages to drilling for oil. It is relatively inexpensive to extract. It is also a reliable and dependable source of energy and money for the local community.

Oil provides us with thousands of conveniences. In the form of gasoline, it is a portable source of energy that gives us the power to drive places. Petroleum is also an ingredient in many items that we depend on.

However, burning gasoline is harmful to the environment. It releases hazardous gases and fumes into the air that we breathe. There is also the possibility of an oil spill. If there is a problem with the drilling machinery, the oil can explode out of the well and spill into the ocean or surrounding land. Oil spills are environmental disasters, especially offshore spills. Oil floats on water, so it can look like food to fish and ruin birds' feathers.

Natural Gas

Natural gas is another fossil fuel that is trapped underground in reservoirs. It is mostly made up of methane. You may have smelled methane before. The decomposing material in landfills also release methane, which smells like rotten eggs.

There is so much natural gas underground that it is measured in million, billion, or trillion cubic meters.

Natural gas is found in deposits a few hundred meters underground. In order to get natural gas out of the ground, companies drill straight down. However, natural gas does not form in big open pockets. Natural gas is trapped in rock formations that can stretch for kilometers.

To reach natural gas, some companies use a process called "hydraulic fracturing," or fracking. *Hydraulic* means they use water, and *fracturing* means to "split apart." The process uses high-pressure water to split apart the rocks underground. This releases the natural gas that is trapped in rock formations. If the rock is too hard, they can send acid down the well to dissolve the rock. They can also use tiny grains of glass or sand to prop open the rock and let the gas escape.

We use natural gas for heating and cooking. Natural gas can also be burned to generate electricity. We rely on natural gas to give power to lights, televisions, air conditioners, and kitchen appliances in our homes.

Natural gas can also be turned into a liquid form, called liquid natural gas (LNG). LNG is much cleaner than any other fossil fuels.

Liquid natural gas takes up much less space than the gaseous form. The amount of natural gas that would fit into a big beach ball would fit into a ping-pong ball as a liquid! LNG can be easily stored and used for different purposes. LNG can even be a replacement for gasoline.

Advantages and Disadvantages

Natural gas is relatively inexpensive to extract, and is a "cleaner" fossil fuel than oil or coal. When natural gas is burned, it only releases carbon dioxide and water vapor (which are the exact same gases that we breathe out when we exhale) This is healthier than burning coal.

However, extracting natural gas can cause environmental problems. Fracturing rocks can cause mini-earthquakes. The high-pressure water and chemicals that are forced underground can also leak to other sources of water. The water sources, used for drinking or bathing, can become contaminated and unsafe.

Other Non-renewable Energy Sources

Fossil fuels are the leading non-renewable energy sources around the world. There are others, however.

Nuclear Energy

Nuclear energy is usually considered another non-renewable energy source. Although nuclear energy itself is a renewable energy source, the material used in nuclear power plants is not.

Nuclear energy harvests the powerful energy in the nucleus, or core, of an atom. Nuclear energy is released through nuclear fission, the process where the nucleus of an atom splits. Nuclear power plants are complex machines that can control nuclear fission to produce electricity.

The material most often used in nuclear power plants is the element uranium. Although uranium is found in rocks all over the world, nuclear power plants usually use a very rare type of uranium, U-235. Uranium is a non-renewable resource.

Nuclear energy is a popular way of generating electricity around the world. Nuclear power plants do not pollute the air or emit greenhouse gases. They can be built in rural or urban areas, and do not destroy the environment around them.

However, nuclear energy is difficult to harvest. Nuclear power plants are very complicated to build and run. Many communities do not have the scientists and engineers to develop a safe and reliable nuclear energy program.

Nuclear energy also produces radioactive material. Radioactive waste can be extremely toxic, causing burns and increasing the risk for cancers, blood diseases, and bone decay among people who are exposed to it.

Biomass Energy

Biomass energy, a renewable energy source, can also be a non-renewable energy source. Biomass energy uses the energy found in plants.

Environmental Impacts of Non-renewable Energy Sources

Greenhouse Gas Emissions

Perhaps the most well-known impact of using non-renewable energy sources is the emission of greenhouse gases, in particular CO_2 and methane, which contribute to climate change.

Different types of non-renewable energy though emit different levels of greenhouse gases. For example, coal is considered the worst to emit the highest percentage of CO_2.

Indeed, natural gas emits a lot less CO_2, specifically 50-60% less compared to coal and it also emits 15-20% fewer heat-trapping gases compared to gasoline when used to power a vehicle.

However, that does not mean that natural gas can help mitigate climate change as drilling and extracting natural gas from wells results in the leakage of methane, which is a much more potent greenhouse gas – it is 34 times stronger than CO_2 in terms of its potential for trapping heat.

The concerns about greenhouse gas emissions and climate change are cross-cutting. It is not just about the direct impact of rising temperatures and changing weather patterns which impact human livelihoods as floods or dry seasons proliferate.

Climate change is impacting ecosystems, diminishing their capacity to adapt to changing climates and as a result threatening biodiversity and the important ecosystem services our lives rely on.

Air Pollution

Non-renewable energy sources are not just altering our Earth's atmosphere by increasing the amount of greenhouse gas emissions. They also emit a variety of pollutants that are impact people's health and the environment.

For example, coal-fired power plants are the single largest source of mercury emissions in the US. When mercury is emitted into the air, it goes on to settle on the ground or blends with water. From that point it accumulates on organisms such as fish, passing through the food chain.

This has profound effects on our biodiversity but creates real risks for people as studies have found that exposure to mercury can lead to neurological and neuro-behavioural effects in embryos and young children.

Other air pollutants emitted due to fossil fuel combustion include sulphur dioxide, nitrogen oxides and particulate matter.

Acid Rain and Water Pollution

But it is not just the air that we breathe that gets polluted. Dangerous pollutants that are emitted into the air can often follow the water cycle. This is the case of acid rain which is created due to the emission of sulphur and other chemicals into the atmosphere. Because of this cocktail of chemicals, the rain becomes mildly acidic.

Acid rain is corrosive to machinery and can disrupt local ecosystems. In terms of the environment, acid rain changes the acidity of lakes and streams which can be very harmful to fish and other aquatic organisms; it would also be damaging to trees thereby weakening forest ecosystems.

Another aspect of water pollution due to the use of fossil fuels such as coal, or nuclear energy is what is known as "thermal pollution".

Both fossil fuel plants and nuclear power plants require water to run and help cool the power plant. The water they use is also needed for other purposes, such as maintaining local ecosystem functions or agriculture. When the plants release that water back into the environment, its temperature is changed and as a result its quality has been degraded.

Heated water that is reintroduced to the environment contains lower levels of dissolved oxygen which can stress native wildlife by for example increasing the heart rate of fish or decreasing fertility.

Land Pollution and Waste Generation

It is also important not to forget environmental impacts that come about as a result of the extraction of non-renewable resources or the disposal of the waste they generate.

There is very clear evidence illustrating the impact of surface mining both in the short and long-term. For instance, huge volumes of excess rock or soil are dumped in other locations such as nearby valleys affecting those ecosystems.

When it comes to the land that is being mined, in the long term these sites are left with poor quality soil and sometimes due to the chemicals used the lands ends up being polluted as well as any nearby water reserves.

The case of nuclear energy is of particular relevant here as both the extraction of uranium and waste disposal create some very critical issues for which no long-term solution has been found due to the radioactive nature of the ore being mined.

Oil Spills and Other Accidents

Lastly, there is something to be said about unintended consequences or rather unforeseen and accidental effects. Oil spills are extremely damaging to nearby shores and ecosystems.

Similarly devastating effects on both nature and humans were seen as a result of the nuclear disaster in Chernobyl. These were accidents which came about due to negligence, technology failure, lack of preparedness or a combination of all those.

Production of Coal

Modern life is unimaginable without electricity. It lights houses, buildings, streets, provides domestic and industrial heat, and powers most equipment used in homes, offices and machinery in factories. Improving access to electricity worldwide is critical to alleviating poverty.

Coal plays a vital role in electricity generation worldwide. Coal-fuelled power plants currently fuel 37% of global electricity and, in some countries, coal fuels a higher percentage of electricity.

Coal fired power plants are a type of power plant that make use of the combustion of coal in order to generate electricity. Their use provides around 40% of the world's electricity and they are primarily used in developing countries. Countries such as South Africa use coal for 94% of their electricity and China and India use coal for 70-75% of their electricity needs, however the *amount* of coal China uses dwarfs most other countries. The use of coal provides access to electricity to those who previously didn't have it, which helps to increase quality of life and reduce poverty in those regions, however it produces large quantities of different pollutants which reduces air quality and contributes to climate change.

Burning Huge Amounts of Coal

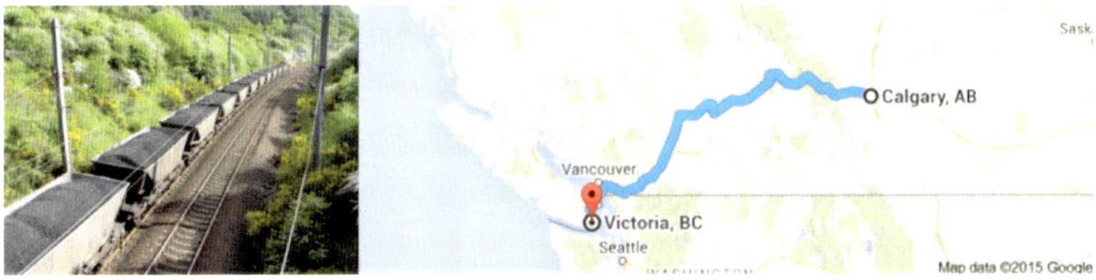

Figure: A coal train the length of 1100 kilometers, the distance from Calgary to Victoria, is needed per year for a 1000 MWe coal fired power plant.

Coal plants require enormous amounts of coal. Shockingly: a 1000 MWe coal plant uses 9000 tonnes of coal per day, equivalent to an entire train load (90 cars with 100 tonnes in each!). The amount of coal used during a full year would then require 365 trains, and if each is 3 km long then a single train carrying all of this coal would need to be about 1100 km long; If this train were to pass by your house at 40 kilometers per hour, it would take more than *a day* to pass.

The conversion of this coal to the end goal of electricity is a multi-faceted process:

1. The coal must be unloaded from the train. Traditional ways of doing this require the use of cranes picking up the coal from the cars, however newer plants have the floor underneath the train tracks drop away, allowing the coal to be dropped into underground containment. Doing so doesn't even require the train to stop moving. Many coal plants are mine mouth which means the plant was put where the coal mine is, so the coal doesn't need to be transported by train.
2. Once unloaded, the coal is then pulverized into a fine powder by a large grinder. This ensures nearly complete burning of the coal in order to maximize the heat given off and to minimize pollutants.
3. The pulverized coal is then input to a boiler, where combustion occurs and the coal provides heat to the power plant. This heat is transferred to pipes containing high pressured water, which boils to steam.
4. The steam then travels through a turbine, causing it to rotate extremely fast which in turn spins a generator, producing electricity. The electricity can then be input to the electrical grid for use by society.

Coal fired power plants follow the Rankine cycle in order to complete this process. Since they require plenty of water to be circulated in this cycle, coal power plants need to be located near a body of water. The process of coal fired plants can be seen below in figure below.

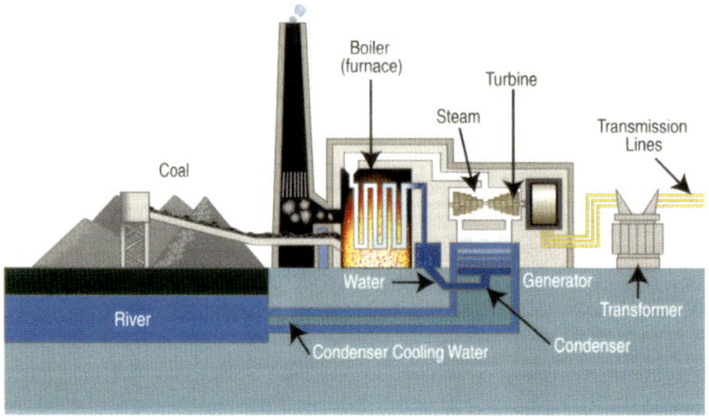

Figure: The process of a coal fired power plant to convert coal into electricity.

Efficiency Improvements

Improvements continue to be made in conventional PCC power station design and new combustion technologies are being developed. These allow more electricity to be produced from less coal - known as improving the thermal efficiency of the power station. Efficiency gains in electricity generation from coal-fired power stations will play a crucial part in reducing CO_2 emissions at a global level. A one percentage point improvement in the efficiency of a conventional pulverised coal combustion plant results in a 2-3% reduction in CO2 emissions.

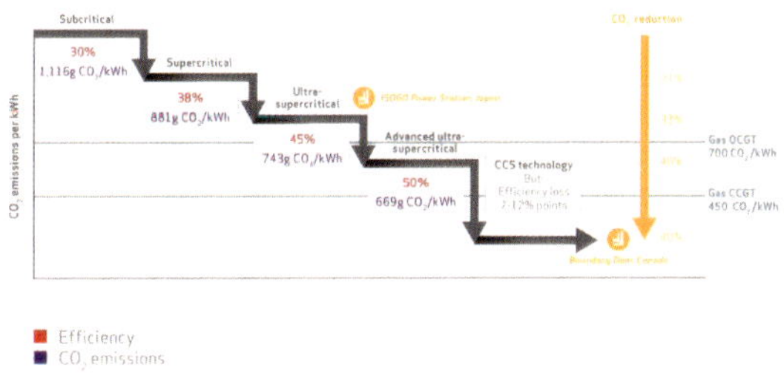

Environmental Impact of Coal

Air Pollution

The smoke from coal power plants is exceedingly dangerous to human health.

When coal is burned it releases a number of airborne toxins and pollutants. They include mercury, lead, sulfur dioxide, nitrogen oxides, particulates, and various other heavy metals. Health impacts can range from asthma and breathing difficulties, to brain damage, heart problems, cancer, neurological disorders, and premature death.

Although limits set by the Environmental Protection Agency (EPA) have helped prevent *some* of these emissions, many plants don't have the necessary pollution controls installed. The future of these protections remains unclear.

Coal Impacts: Water Pollution

Acid rock drainage from a mine in Ohio.

When you burn charcoal in your grill at home, ash is leftover. The same is true for coal-fired power plants, which produce more than 100 million tons of coal ash *every year*. More than half of that waste ends up in ponds, lakes, landfills, and other sites where, over time, it can contaminate waterways and drinking water supplies.

Other water impacts include acid rock drainage from coal mines, the obliteration of mountain streams and valleys by mountain top removal mining, and the energy-water collisions that occur when coal plants rely too heavily on local water supplies.

Coal impacts: Global Warming

Coal causes global warming, which is linked with extreme weather and wildfires.

Climate change is coal's most serious, long-term, global impact. Chemically, coal is *mostly* carbon, which, when burned, reacts with oxygen in the air to produce carbon dioxide, a heat-trapping gas. When released into the atmosphere, carbon dioxide works like a blanket, warming the earth above normal limits.

Consequences of global warming include drought, sea level rise, flooding, extreme weather, and species loss. The severity of those impacts is tied directly to the amount of carbon dioxide we release, including from coal plants.

Production of Natural Gas

Natural gas is mainly composed of methane (CH_4) and is a very potent greenhouse gas. There are two types of natural gas. Biogenic gas is found at shallow depths and arises from anaerobic decay of organic matter by bacteria, like landfill gas. Thermogenic gas comes from the compression of organic matter and deep heat underground. They are found with petroleum in reservoir rocks and with coal deposits, and these fossil fuels are extracted together.

Natural gas is released into the atmosphere from coal mines, oil and gas wells, and natural gas storage tanks, pipelines, and processing plants. When natural gas is produced but cannot be captured and transported economically, it is "flared," or burned at well sites, which converts it to CO_2. This is considered to be safer and better than releasing methane into the atmosphere because CO_2 is a less potent greenhouse gas than methane.

Natural gas is a preferred fossil fuel when considering its environmental impacts. Specifically, when burned, much less carbon dioxide (CO_2), nitrogen oxides, and sulfur dioxide are omitted than from the combustion of coal or oil. It also does not produce ash or toxic emissions.

Natural gas production can result in the production of large volumes of contaminated water. This water has to be properly handled, stored, and treated so that it does not pollute land and water supplies. Extraction of shale gas is more problematic than traditional sources due to a process nicknamed fracking, or fracturing of wells, since it requires large amounts of water. The technique uses high-pressure fluids to fracture the normally hard shale deposits and release gas and oil trapped inside the rock. To promote the flow of gas out of the rock, small particles of solids are included in the fracturing liquids to lodge in the shale cracks and keep them open after the liquids are depressurized. The considerable use of water may affect the availability of water for other uses in some regions and this can affect aquatic habitats. If mismanaged, hydraulic fracturing fluid can be released by spills, leaks, or various other exposure pathways. The fluid contains potentially hazardous chemicals such as hydrochloric acid, glutaraldehyde, petroleum distillate, and ethylene glycol.

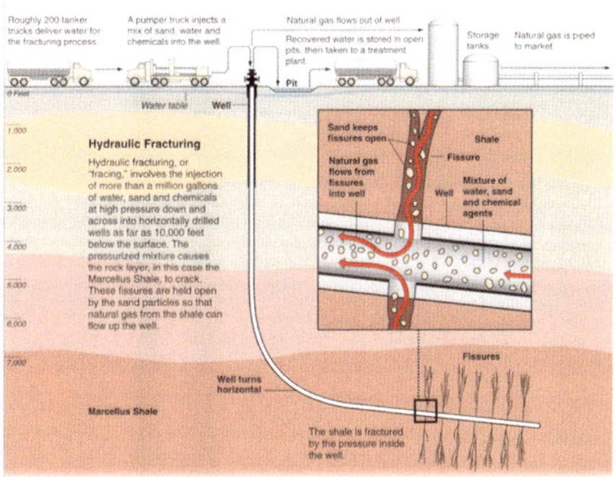

Figure: Graphic illustrates the process of hydraulic fracturing.

The raw gas from a well may contain many other compounds besides the methane that is being sought, including hydrogen sulfide, a very toxic gas. Natural gas with high concentrations of hydrogen sulfide is usually flared which produces CO_2, carbon monoxide, sulfur dioxide, nitrogen oxides, and many other compounds. Natural gas wells and pipelines often have engines to run equipment and compressors, which produce additional air pollutants and noise.

Steam Generation Units

Natural gas can be used to generate electricity in a variety of ways. The most basic natural gas fired electric generation consists of a steam generation unit, where fossil fuels are burned in a boiler to heat water and produce steam that then turns a turbine to generate electricity. Natural gas may be used for this process, although these basic steam units are more typical of large coal or nuclear generation facilities. These basic steam generation units have fairly low energy efficiency. Typically, only 33 to 35 percent of the thermal energy used to generate the steam is converted into electrical energy in these types of units.

Centralized Gas Turbines

Gas turbines and combustion engines are also used to generate electricity. In these types of units, instead of heating steam to turn a turbine, hot gases from burning fossil fuels (particularly natural gas) are used to turn the turbine and generate electricity. Gas turbine and combustion engine plants are traditionally used primarily for peak-load demands, as it is possible to quickly and easily turn them on. These plants have increased in popularity due to advances in technology and the availability of natural gas. However, they are still traditionally slightly less efficient than large steam-driven power plants.

A Centralized Gas Turbine Generation Station

Turbo-gas Plants

Natural gas can also be used in turbo-gas electric plants. These thermoelectric plants directly exploit the energy produced during methane (or gas oil) combustion and work without boiler, in order to transform the water into vapor and without condenser in order to reconvert the vapor into water. A turbo-gas plant consists of:

- A compressor: It sucks air up from the atmosphere, compresses it and sends it to the combustion chamber.
- Combustion chamber: This is where the combustion between the air and the fuel occurs (methane or gas oil).
- Gas turbine: the air and gas mixture, at high temperature, gets into a turbine where the expansion of combusted gases makes the rotor blades rotate and subsequently activates the alternator therefore generating electricity. The advantages of turbo-gas plants are: the low costs of the plant, they start rapidity even if there is no energy in the network and the fact that these plants do not need cooling water. It is possible to build them in any place, even far from the rivers and the sea. The disadvantage is the very low output (around 30%) and therefore the very high energy cost.

Combined-cycle Plants

Combined-cycle and co-generation systems are the most efficient technologies to produce electricity from natural gas. Both use the heat that would normally be lost. Combined-cycle plants exploit the heat they generate to produce electricity. These systems associate a turbo-gas plant with a vapour group: the residual heat of fumes going out from the turbo-gas group is used to produce vapour, increasing the performance by 56%. Moreover combined-cycle plants have lower building and maintenance costs, and a higher functioning reliability. Combined cycle plants represent the most sustainable solution for thermoelectric power generation, guaranteeing a low overall impact on natural resources. In particular, the use of natural gas has a very limited environmental impact in terms of local atmospheric pollution. Respect to coal and oil, in fact, using natural gas eliminates sulphur dioxide and dust emissions and reduces nitrogen oxide emissions thanks to the introduction of state-of-the-art combustion technology. In terms of global atmospheric pollution, the combination natural gas/combined cycle represents one of the most efficient alternatives to reduce greenhouse gas emissions. Natural gas is the fossil fuel whose combustion to produce a unit amount of energy generates the smallest amount of carbon dioxide: the "emission factor" amounts to 2.35 tCO_2/tep, around 26% lower that of petroleum-based products and around 41% lower than coal.

Distributed Generation

A Proposed Natural Gas Combined Cycle Power Plant In New York

Until recently, methods of generating power have been discussed in the context of large, centralized power plants. However, with technological advancements, there is a trend towards what is known as 'distributed generation'. Distributed generation refers to the placement of individual, smaller sized electric generation units at residential, commercial, and industrial sites of use. These small scale power plants, which are primarily powered by natural gas, operate with small gas turbine or combustion engine units, or natural gas fuel cells.

Distributed generation can take many forms, from small, low output generators used to back up the supply of electricity obtained from the centralized electric utilities, to larger, independent generators that supply enough electricity to power an entire factory. Distributed generation is attractive because it offers electricity that is more reliable, more efficient, and cheaper than purchasing power from a centralized utility. Distributed generation also allows for increased local control over the electricity supply, and cuts down on electricity losses during transmission.

Natural gas is one of the leading energy sources for distributed generation. Because of the extensive natural gas supply infrastructure and the environmental benefits of using natural gas, it is one of the leading choices for on-site power generation. There are a number of ways in which natural gas may be used on-site to generate electricity. Fuel cells, gas-fired reciprocating engines, industrial natural gas-fired turbines, and microturbines are all popular forms of using natural gas for on-site electricity needs.

Industrial Natural Gas Fired Turbines

Industrial natural gas-fired turbines operate on the same concept as the larger centralized gas turbine generators discussed above. However, instead of being located in a centralized plant, these turbines are located in close proximity to where the electricity being generated will be used. Industrial turbines – producing electricity through the use of high temperature, high pressure gas to turn a turbine that generates a current – are compact, lightweight, easily started, and simple to operate. This type of distributed generation is commonly used by medium and large sized establishments, such as universities, hospitals, commercial buildings and industrial plants, and can achieve efficiency up to 58 percent.

In contrast with distributed generation the heat that would normally be lost as waste energy can easily be harnessed to perform other functions, such as powering a boiler or space heating. This is known as Combined Heat and Power (CHP) systems. These systems make use of heat that is normally wasted in the electric generation process, thereby increasing the energy efficiency of the total system.

Microturbines

Microturbines are scaled down versions of industrial gas turbines. As their name suggests, these generating units are very small, and typically have a relatively small electric output. These types of distributed generation systems have the capacity to produce from 25 to 500 kilowatts (kW) of electricity, and are best suited for residential or small scale commercial units.

Advantages to microturbines include a very compact size (about the same size as a refrigerator), a small number of moving parts, light-weight, low-cost, and increased efficiency. Using new waste heat recovery techniques, microturbines can achieve energy efficiencies of up to 80 percent.

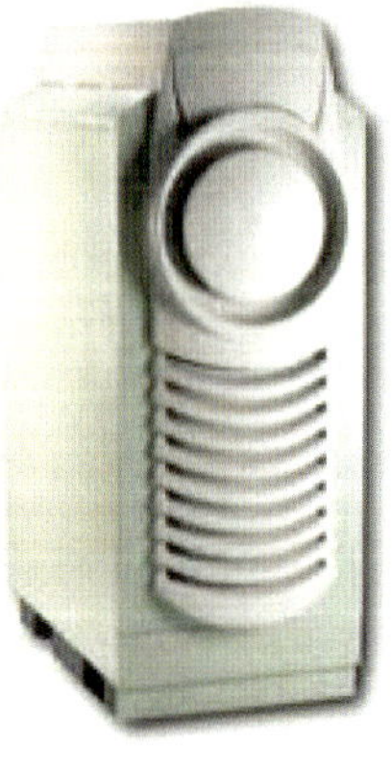

Gas Fired Microturbine

Natural Gas-Fired Reciprocating Engines

Natural-gas fired reciprocating engines are also used for on-site electric generation. These types of engines are also commonly known as combustion engines. They convert the energy contained in fossil fuels into mechanical energy, which rotates a piston to generate electricity. Natural-gas fired reciprocating engines typically generate from less than 5 kW, up to 7 megawatts (MW), meaning they can be used as a small scale residential backup generator to a base load generator in industrial settings. These engines offer efficiencies from 25 to 45 percent, and can also be used in a CHP system to increase energy efficiency.

Gas Fired Reciprocating Engine

Fuel cells are becoming an increasingly important technology for the generation of electricity. They are much like rechargeable batteries, except instead of using an electric recharger, they use a fuel, such as natural gas, to generate electric power even when they are in use. Fuel cells for distributed generation offer a multitude of benefits, and are an exciting area of innovation and research for distributed generation applications.

Production of Oil

Oil, otherwise known as 'petroleum or crude oil', is a thick black liquid composed primarily of hydrogen and carbon. Oil also contains trace elements of sulphur, nitrogen and oxygen.

Today's oil deposits were formed millions of years ago, when dead marine organisms sunk to the bottom of the ocean and were buried under deposits of sedimentary rock. Subject to intense heat and pressure, these organisms underwent a transformation in which they were converted to oil over millions of years. This process is similar to the process of natural gas formation, however oil forms under a limited range of temperatures while natural gas forms under a wider range. This limited range of temperatures is called the 'oil window'.

Oil is found in specific underground rocks called reservoirs. The rocks have tiny spaces in them that allow them to hold water, natural gas and/or oil. Impermeable rocks called cap rocks surround the reservoir and trap oil in its place.

Through exploratory activities such as seismic, well sampling, and subsurface mapping, geoscientists locate sites for oil drilling.

Oil is extracted from the reservoir by drilling a well and pumping it up the well. Once recovered, oil is transported by pipeline, ship, rail, or truck to a refinery where it undergoes a complex process that produces petroleum products such as gasoline, diesel, jet-fuel, home-heating fuel, lubricating oil and asphalt along with petrochemicals that are used to make common products such as plastic, drugs, synthetic fiber, soap and paint.

Ultimately all power stations created electricity in the same way. When using oil, the oil is burned to heat the water to produce steam. This steam turns a turbine which then spins a set of wire loops exposed to a magnetic field which causes an electrical flow in the wire, producing electricity.

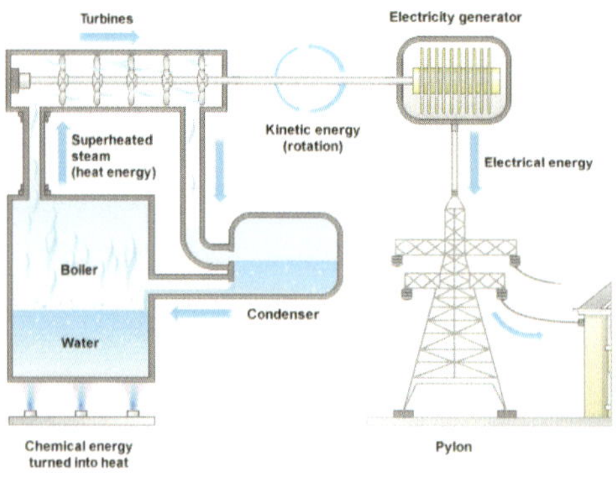

Environmental Impact

Burning oil to generate electricity produces significant air pollution in the forms of nitrogen oxides, and, depending on the sulfur content of the oil, sulfur dioxide and particulates. Carbon dioxide and methane (as well as other greenhouse gases), heavy metals such as mercury, and volatile organic compounds (which contribute to ground-level ozone) all can come out of the smoke stack of an oil-burning power plant.

The operation of oil-fired power plants also impacts water, land use and solid waste disposal. Similar to the operations of other conventional steam technologies, oil-fired conventional steam plants require large amounts of water for steam and cooling, and can negatively impact local water resources and aquatic habitats. Sludges and oil residues that are not consumed during combustion became a solid waste burden and contain toxic and hazardous wastes.

Drilling also produces a long list of air pollutants, toxic and hazardous materials, and emissions of hydrogen sulfide, a highly flammable and toxic gas. All of these emissions can impact the health and safety of workers and wildlife. Loss of huge stretches of wildlife habitat also occur during drilling. Refineries, too, spew pollution into the air, water and land (in the form of hazardous wastes). Oil transportation accidents can result in catastrophic damage killing thousands of fish, birds, other wildlife, plants and soil.

Production of Nuclear Energy

A nuclear reactor produces and controls the release of energy from splitting the atoms of uranium.

Uranium-fuelled nuclear power is a clean and efficient way of boiling water to make steam which drives turbine generators. Except for the reactor itself, a nuclear power station works like most coal or gas-fired power stations.

Reactor Core

Several hundred fuel assemblies containing thousands of small pellets of ceramic uranium oxide fuel make up the core of a reactor. For a reactor with an output of 1000 megawatts (MWe), the core would contain about 75 tonnes of enriched uranium.

In the reactor core, the uranium-235 isotope fissions or splits, producing a lot of heat in a continuous process called a chain reaction. The process depends on the presence of a moderator such as water or graphite, and is fully controlled.

The moderator slows down the neutrons produced by fission of the uranium nuclei so that they go on to produce more fissions.

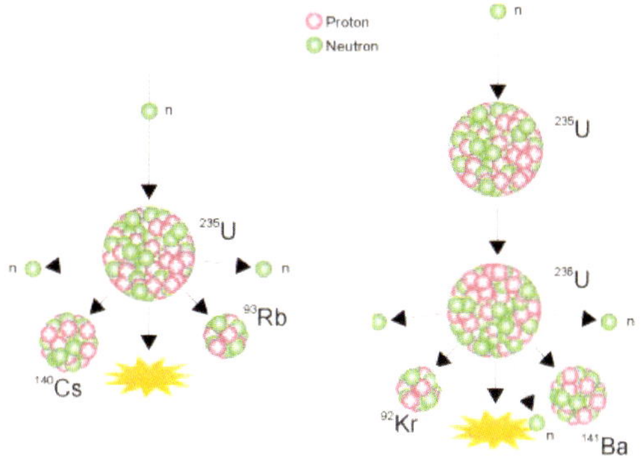

Two examples of fission of a uranium-235 atom

Some of the uranium-238 in the reactor core is turned into plutonium and about half of this is also fissioned similarly, providing about one third of the reactor's energy output.

The fission products remain in the ceramic fuel and undergo radioactive decay, releasing a bit more heat. They are the main wastes from the process.

The reactor core sits inside a steel pressure vessel, so that water around it remains liquid even at the operating temperature of over 320°C. Steam is formed either above the reactor core or in separate pressure vessels, and this drives the turbine to produce electricity. The steam is then condensed and the water recycled.

Common Types of Reactor

The main design is the pressurised water reactor which has water in its primary cooling/heat transfer circuit, and generates steam in a secondary circuit. The less popular boiling water reactor makes steam in the primary circuit above the reactor core, though it is still under considerable pressure. Both types use water as both coolant and moderator, to slow neutrons.

To maintain efficient reactor performance, about one-third or half of the used fuel is removed every year or two, to be replaced with fresh fuel.

The pressure vessel and any steam generators are housed in a massive containment structure with reinforced concrete about 1.2 meters thick. This is to protect neighbors if there is a major problem inside the reactor, and to protect the reactor from external assaults.

Because some heat is generated from radioactive decay even after the reactor is shut down, cooling systems are provided to remove this heat as well as the main operational heat output.

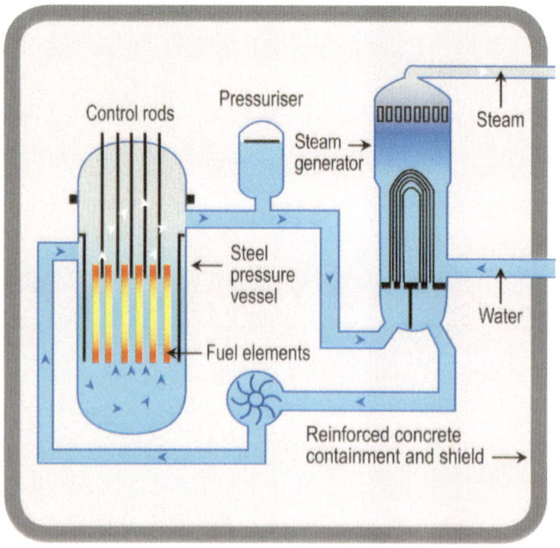

Diagram of pressurised water reactor

Effects of Nuclear Power Plants on Environment

Though the Nuclear Energy Power Plants (NEPS) that leaves its mark of "atom era" on this century is a clear, reliable and settled technology in electric production, it takes reactions by the public in many countries.

The effects of nuclear plants on environment appear during taking out of uranium and thorium, preparation of fuel, production, enriching, retreatment of fuel, storing and detaching of reactors. The biggest effect of nuclear plants on environment is emission of a radioactive matter in environment as a result of an accident. Gases and liquid radioactive wastes from nuclear plants cause significant environmental effects. However, the effects of radiation on environment vary depending on power of accident, type of reactor and security system out of reactor. If various radioisotopes disperse to environment as a result of the accident, radiation contaminated to water, soil and air taking medium effects the environment and human health. Here the important thing is that well-conditioned storing and keeping of high level radiating wastes after the fuel completes its usage life.

The radioactive effects reach to environment and all living beings including humans by means of two different ways. The first way: transportation of emissions arising from funnels in the atmosphere and their reaching to the earth and living beings on earth. The second way: reaching of liquid and solid wastes arising from power plants to rivers, lakes and seas and their effect on living beings and underground waters. Because of the circulation of natural life, the human beings and animals living on earth can effect from the radioactivity arising from nuclear power plants by means of both ways.

Production of Renewable Energy

Renewable energy is energy that comes from sources that replenish themselves over short periods of time. For the most part, renewable energy sources also provide clean energy, or energy that emits few greenhouse gases or pollutants. For this reason, many policy experts and scientists advocate renewable energy sources over traditional fossil fuels. The difficulty is achieving the technology, infrastructure, and political support to make this transition.

The five renewable energy sources highlighted in this map series are the five largest worldwide.

Hydro

Electrical energy created from water stored in huge dams. The energy created by the water released from these dams is transformed into electricity by hydro-electric turbines and generators. The most famous source of hydroelectric power is in the Snowy Mountains, NSW. It is less expensive than mining fossil fuels and does not contribute to the greenhouse effect.

Sun

By generating electricity through the sun we prevent the release into the atmosphere of around 500 tonnes of greenhouse gases each year.

Wind

Moving air which is created when the sun heats the air and cooler air moves in to replace it. This causes wind. Through the ages people have learned to harness the wind's energy. Like the sun, it can also be used to create electricity. Wind power generates less than 1% of the electricity in Australia, but more wind farms are being built every year.

Biomass

Energy which comes from landfill or rubbish dumps. It includes energy from both animal and plant matter. Landfill gas is created when the waste you throw away starts rotting (or decomposing) in the ground. This gas would normally just seep through the ground and into the atmosphere, contributing to environmental problems, like the greenhouse effect. However, it can be captured and processed to create electricity. It is collected, dried (to get rid of any water), and then filtered (to get rid of any waste particles). It is then fed through pipes to a gas generator that burns the gas to create electricity.

Geothermal

Energy from the heat of the earth. It has been used for thousands of years in some countries for hot water, cooking and heating. It can also generate electricity using steam produced from heat found beneath the surface of the earth.

Production of Geothermal Energy

Geothermal power plants use steam produced from reservoirs of hot water found a few miles or more below the Earth's surface to produce electricity. The steam rotates a turbine that activates a generator, which produces electricity. There are three types of geothermal power plants: dry steam, flash steam, and binary cycle.

This geothermal power plant generates electricity for the Imperial Valley in California.

Dry Steam

Dry steam power plants draw from underground resources of steam. The steam is piped directly from underground wells to the power plant where it is directed into a turbine/generator unit.

Flash Steam

Flash steam power plants are the most common and use geothermal reservoirs of water with temperatures greater than 360° F (182° C). This very hot water flows up through wells in the ground under its own pressure. As it flows upward, the pressure decreases and some of the hot water boils into steam. The steam is then separated from the water and used to power a turbine/generator. Any leftover water and condensed steam are injected back into the reservoir, making this a sustainable resource.

Binary Steam

Binary cycle power plants operate on water at lower temperatures of about 225-360° F (107-182° C). Binary cycle plants use the heat from the hot water to boil a working fluid, usually an organic compound with a low boiling point. The working fluid is vaporized in a heat exchanger and used to turn a turbine. The water is then injected back into the ground to be reheated. The water and the working fluid are kept separated during the whole process, so there are little or no air emissions.

Currently, two types of geothermal resources can be used in binary cycle power plants to generate electricity: enhanced geothermal systems (EGS) and low-temperature or co-produced resources.

Environmental Impacts

Geothermal power plants use fluids drawn from the deep earth. These fluids can carry a mixture of gases, notably carbon dioxide (CO_2), hydrogen sulfide (H_2S), methane (CH_4) and ammonia (NH_3). These pollutants contribute to global warming, acid rain, and noxious smells if released. Existing geothermal electric plants emit an average of 122 kg of CO_2 per megawatt-hour (MWh) of electricity, a small fraction of the emission intensity of conventional fossil fuel plants. Plants that experience high levels of acids and volatile chemicals are usually equipped with emission-control systems to reduce the exhaust. Geothermal plants could theoretically inject these gases back into the earth, as a form of carbon capture and storage.

Sociopolitical Influences

Geothermal power plants improve the standard of living. Because of the need for workers, they improve the education of local people, and because of the working power plant, health and sanitary issues are taken care of. But, there are also negative impacts, as well as possible impacts on drinking water, there are cultural heritage impacts, impacts on the landscape, the soil and rocks, and there are disturbances caused by noise.

Impacts on the Electricity Network

These plants have very good basic strength, capacitance and energy. This means that they produce consistent and well-compensated energy. Unfortunately, the compensated part cannot be used typically because it is needed for the purpose of restoring fluid into the ground.

Production of Solar Energy

Solar energy is produced when sunlight strikes solar panels, which then turn solar power into usable electricity. This photovoltaic transformation is the way solar energy is generated.

Solar panels convert Sunlight to Energy

The process of creating power from sunlight starts with the biggest part of a solar installation: the solar panels. A typical solar panel is made from either monocrystalline or polycrystalline silicon housed in a metal panel frame with a glass casing.

When sunlight strikes the thin layer of silicon on the top of a solar panel, it knocks electrons off the silicon atoms. Electrons are negatively charged, which means they are attracted to one side of the silicon cell. This creates an electric current that is captured by the wiring in a solar panel.

Known as the "photovoltaic effect", flowing electrons that have been knocked free by sunlight are the basis of the electricity generated from solar panels. The more sun that strikes your solar panels, the more electrons will be knocked free, so it is important to consider factors like shade and local cloud cover when designing a solar energy system.

A Silfab polycrystalline solar panel

Different solar panels produce electricity at different rates depending on their efficiency and quality. High quality solar panels will often cost slightly more money up front, but in the long term will save you money because they produce electricity more efficiently.

Inverters Turn Energy from Solar Panels into Usable Electricity

Once loose electrons have been collected at individual panels, the resulting current is known as direct current, or DC. The electricity we use to power devices, charge electric vehicles, and turn on our lights is alternating current, or AC. Electricity generated by loose silicon electrons is DC electricity and needs to be converted to AC electricity for before we can use it.

This conversion is done by important devices called inverters. Inverters can be configured as string inverters, microinverters, or power optimizers, but each setup performs the same general function of turning DC electricity to AC electricity. Once the electricity passes through inverters and is changed to AC electricity, it flows to your electric panel and meter. Finally, it can be dispersed throughout your home or into the electric grid.

Other Solar Panel Technologies

Most solar panels available are made from crystalline silicon, but there are a few other types of solar panels that produce electricity from the sun.

Cadmium Telluride (CdTe) Solar Panels

CdTe panels use a different material than silicon to produce electricity. While they don't have very high efficiency ratings (yet), CdTe panels can capture energy at shorter wavelengths than silicon panels. Additionally, manufacturing costs are low.

Copper Gallium Indium Diselenide (CIGS) Solar Panels

With higher efficiencies than CdTe panels, CIGS panels are a promising panel option. Unfortunately, high production costs have prevented CIGS panels from making it to the mainstream. CIGS panels use a different, copper-based material to generate electricity.

Organic Photovoltaic Cells (OPV)

Instead of using an inorganic material like silicon or copper, OPV cells use thin layers of organic vapor or solution to generate an electrical current from sunlight. OPV cells have yet to see widespread adoption due to both low efficiencies and short life spans when compared to other panel technologies.

These solar technologies all fall under the umbrella of thin film panels, and are a very small part of the overall solar market. This is mainly because they aren't cost-effective at this time. In the future these unique options may be the dominant solar technologies, but for now silicon panels dominate the industry.

Environmental Impacts of Solar Power

Photovoltaics is now a proven technology which is inherently safe, as opposed to some dangerous electricity generating technologies. Over its estimated life a photovoltaic module will produce much more electricity than was used in its production. A 100 W module will prevent the emission of over two tonnes of CO_2. Photovoltaic systems make no noise and cause no pollution while in operation. PV cell technologies that have relatively lower environmental risks compared to other types of electric sources. However, chemicals used in PV cells could be released to air, surface water, and groundwater in the manufacturing facility, the installation site, and the disposal or recycling facility. The production of photovoltaic devices involves the use of a variety of chemicals and materials. The amounts and types of chemicals used will vary depending upon the type of cell being produced. The releases of chemicals to the air from the photovoltaic facilities were reported as both air stack emissions and fugitive air emissions. All six of the photovoltaic companies reviewed reported air stack emissions of some compounds used for cleaning and etching. The chemicals released in the largest quantities in air stack emissions included 1,1,1-trichloroethane, acetone, ammonia, isopropyl alcohol, and methanol. The scale of the system plays a significant role in the level of environmental impact. Depending on their location, larger utility-scale solar facilities can raise concerns about land degradation and habitat loss and impacts from utility-scale solar systems can be minimized by siting them at lower-quality locations such as abandoned mining land, or existing transportation and transmission corridors.

Solar PV cells do not use water for generating electricity. However, as in all manufacturing processes, some water is used to manufacture solar PV components. Concentrating solar thermal plants (CSP), like all thermal electric plants, require water for cooling. Water use depends on the plant design, plant location, and the type of cooling system.

Production of Wind Energy

Wind is a form of solar energy. Winds are caused by the uneven heating of the atmosphere by the sun, the irregularities of the earth's surface, and rotation of the earth. Wind flow patterns are modified by the earth's terrain, bodies of water, and vegetative cover. This wind flow, or motion energy, when "harvested" by modern wind turbines, can be used to generate electricity.

Generation of Wind Power

The terms "wind energy" or "wind power" describe the process by which the wind is used to generate mechanical power or electricity. Wind turbines convert the kinetic energy in the wind into mechanical power. This mechanical power can be used for specific tasks (such as grinding grain or pumping water) or a generator can convert this mechanical power into electricity to power homes, businesses, schools, and the like.

Wind Turbines

Wind turbines, like aircraft propeller blades, turn in the moving air and power an electric generator that supplies an electric current. Simply stated, a wind turbine is the opposite of a fan. Instead of using electricity to make wind, like a fan, wind turbines use wind to make electricity. The wind turns the blades, which spin a shaft, which connects to a generator and makes electricity.

Types of Wind Turbine

Modern wind turbines fall into two basic groups; the horizontal-axis variety, like the traditional farm windmills used for pumping water, and the vertical-axis design, like the eggbeater-style Darrieus model, named after its French inventor. Most large modern wind turbines are horizontal-axis turbines.

Turbine Components

Horizontal turbine components include:

- Blade or rotor, which converts the energy in the wind to rotational shaft energy;
- A drive train, usually including a gearbox and a generator;
- A tower that supports the rotor and drive train; and
- Other equipment, including controls, electrical cables, ground support equipment, and interconnection equipment.

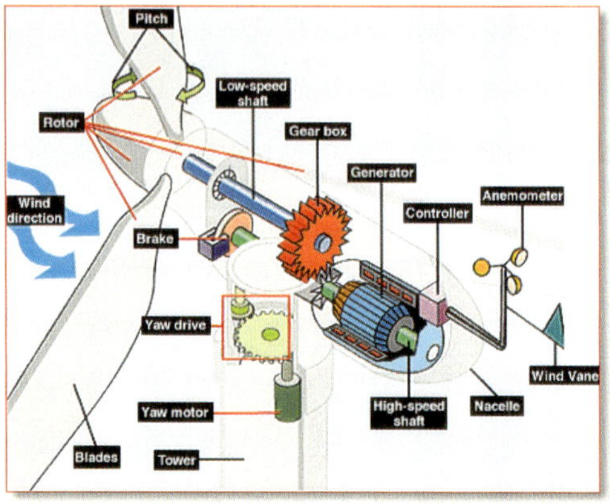

Turbine Configurations

Wind turbines are often grouped together into a single wind power plant, also known as a wind farm, and generate bulk electrical power. Electricity from these turbines is fed into a utility grid and distributed to customers, just as with conventional power plants.

Wind Turbine Size and Power Ratings

Wind turbines are available in a variety of sizes, and therefore power ratings. The largest machine has blades that span more than the length of a football field, stands 20 building stories high, and produces enough electricity to power 1,400 homes. A small home-sized wind machine has rotors between 8 and 25 feet in diameter and stands upwards of 30 feet and can supply the power needs of an all-electric home or small business. Utility-scale turbines range in size from 50 to 750 kilowatts. Single small turbines, below 50 kilowatts, are used for homes, telecommunications dishes, or water pumping.

Advantages and Disadvantages of Wind-Generated Electricity

A Renewable Non-Polluting Resource

Wind energy is a free, renewable resource, so no matter how much is used today, there will still be the same supply in the future. Wind energy is also a source of clean, non-polluting, electricity. Unlike conventional power plants, wind plants emit no air pollutants or greenhouse gases. According to the U.S. Department of Energy, in 1990, California's wind power plants offset the emission of more than 2.5 billion pounds of carbon dioxide, and 15 million pounds of other pollutants that would have otherwise been produced. It would take a forest of 90 million to 175 million trees to provide the same air quality.

Cost Issues

Even though the cost of wind power has decreased dramatically in the past 10 years, the technology requires a higher initial investment than fossil-fueled generators. Roughly 80% of the cost is the machinery, with the balance being site preparation and installation. If wind generating systems are compared with fossil-fueled systems on a "life-cycle" cost basis (counting fuel and operating expenses for the life of the generator), however, wind costs are much more competitive with other generating technologies because there is no fuel to purchase and minimal operating expenses.

Environmental Concerns

Although wind power plants have relatively little impact on the environment compared to fossil fuel power plants, there is some concern over the noise produced by the rotor blades, aesthetic (visual) impacts, and birds and bats having been killed (avian/bat mortality) by flying into the rotors. Most of these problems have been resolved or greatly reduced through technological development or by properly siting wind plants.

Supply and Transport Issues

The major challenge to using wind as a source of power is that it is intermittent and does not always blow when electricity is needed. Wind cannot be stored (although wind-generated electricity can be stored, if batteries are used), and not all winds can be harnessed to meet the timing of electricity demands. Further, good wind sites are often located in remote locations far from areas of electric power demand (such as cities). Finally, wind resource development may compete with other uses for the land, and those alternative uses may be more highly valued than electricity generation. However, wind turbines can be located on land that is also used for grazing or even farming.

Environmental Impacts of Wind Power

A wind farm, when installed on agricultural land, has one of the lowest environmental impacts of all energy sources: it occupies less land area per kilowatt-hour (kWh) of electricity generated than any other energy conversion system, apart from rooftop solar energy, and is compatible with grazing and crops; it generates the energy used in its construction in just 3 months of operation, yet its operational lifetime is 20– 25 years; greenhouse gas emissions and air pollution produced by its construction are very tiny and declining. There are no emissions or pollution produced by its operation; in substituting for base-load (mostly coal power) wind power produces a net decrease in greenhouse gas emissions and air pollution, and a net increase in biodiversity; modern wind turbines are almost silent and rotate so slowly (in terms of revolutions per minute) that they are rarely a hazard to birds.

Modern wind turbine designs have significantly reduced the noise from turbines. Turbine designers are working to minimize noise, as noise reflects lost energy and output. Noise levels at nearby residences are managed through the siting of turbines, the approvals process for wind farms and operational management of the wind farm. The noise limit for wind farms is 35 A-weighted decibels, which is usually around 5 A weighted decibels above a quiet countryside. Alternatively, the limit is 5 A-weighted decibels above the level of background noise (i.e. without wind farm noise), if that is greater than 35 A-weighted decibels. Low frequency sound and infrasound (i.e. usually beneath the threshold of human hearing) are everywhere in the environment. They are emitted from natural sources such as wind and rivers and artificial sources such as traffic and air conditioning. Modern turbine designs which locate the blades upwind instead of downwind have significantly reduced the level of infrasound. Scientific and health authorities have found the low level of infrasound emitted by wind turbines pose no health risks.

Wind turbines may create shadow flicker on nearby residences when the sun passes behind the turbine. However, this can easily be avoided by locating the wind farm to avoid unacceptable shadow flicker, or turning the turbine off for the few minutes of the day when the sun is at the angle that causes flicker. Shadow flicker is considered in the NSW development assessment process to ensure potential impacts are addressed. Many energy policy studies have noted how wind turbines present direct and indirect hazards to birds, other avian species, and bats. Birds can directly smash into moving or even stationary turbine blades, crash into towers and nacelles, and collide with local distribution lines. These risks are exacerbated when turbines are placed on ridges and upwind slopes or built close to migration routes. Some species, such as bats, face additional risks from the rapid reduction in air pressure near turbine blades, which can cause internal hemorrhaging.

Comparative assessment of avian mortality for fossil fuel, nuclear, and wind power plants in the United States is presented in. It was stated that for wind turbines, the risk appears to be greatest to birds striking towers or turbine blades. For fossil-fueled power stations, the most significant fatalities come from climate change, which is altering weather patterns and destroying habitats that birds depend on. For nuclear power plants, the risk is almost equally spread across hazardous pollution at uranium mine sites and collisions with draft cooling structures. Yet, taken together, fossil-fueled facilities are about 17 times more dangerous to birds on a per GWh basis than wind and nuclear power stations.

Production of Hydropower Energy

Hydroelectric power is a renewable energy source which harnesses the power of moving water to produce electricity. The hydroelectric process starts long before you turn your light on at home or work.

Fast moving or falling water is the primary ingredient needed to create hydroelectric power. In most large scale constructions, the water is held back behind a dam and the force of the water being released is used to drive the blades of a turbine which in turn creates electricity that can be fed into the grid. We are also looking at new and innovative ways to transform the energy of moving water into power for our homes through the tides of our oceans and fast flowing rivers.

Large Scale Hydroelectricity

Most power plants that use water share the same basic components that all lead to a valuable source of renewable energy.

- A reservoir or dam: You'll often find people sailing on it or swimming in the water as many reservoirs are used for recreational purposes but they have a far more important purpose. A dam allows the power plant to control the amount of water that is fed into the system and helps to maintain the constant supply of electricity.
- The pipeline or penstock: This is a channel that leads downwards towards the power plant turbine and is usually protected by a gate that can be opened and closed to control the amount of water coming through – it gradually pressurises the water as it approaches the turbine.
- The turbine: Most power stations use the Francis Turbine, a large disc with curved blades which turn as the water hits them. This mechanical energy is then used to turn a central shaft. A turbine can be over 170 tonnes in weight and turn at a speed of nearly a 100 revolutions each minute.
- The generator: The heart of electricity production is a series of magnets which are turned by the shaft of the turbine, causing the production of alternating current or AC.
- The transformer: This alternating current is then fed into a transformer that is converted into a higher voltage which is then fed into power lines.

The water from the operation can either flow down the dam to a river continuing on its way or it can be collected in a secondary reservoir and the water can be pumped back up to be used again.

Smaller Scale Hydroelectric Power Systems

Anyone who has ever been white water rafting or stood on a surf board will be aware of the true power that moving water generates. In recent years we have started to develop more localized applications for hydroelectric power that means we now have the technology to provide energy for homes and commercial premises in a similar way that we do for solar and wind.

The system works the same as for the major power plants except that the water is normally from a localized source like a fast flowing river. A standard hydro system will contain the following components:

An intake, through which the water flows (this has a filter which stops debris and fish from entering and damaging the hydro system, and vice versa).

- A penstock or pressurized pipe between the water intake and turbine through which the water drops.
- A turbine which converts the pressure of the water into mechanical energy (this can be located either in a powerhouse or, in the case of barrage systems, in the river itself).
- Control panels which monitor the system's output.
- A generator that converts the rotational energy from the turbine shaft into electricity.
- An export power cable that connects the system to buildings and/or to the electricity grid.
- A generation (and export) meter which is needed if you are going to claim Feed-in tariffs. This needs to be a Microgeneration Certificated Scheme approved meter.

- A powerhouse which will house the turbine, generator and meter.
- A tailrace or canal from the powerhouse through which the water returns to the river.
- In the case of off-grid systems, a battery to store the energy generated.

Environmental Impact

The physical environment is affected rather significantly by the construction of a hydroelectric power station. Both the river and ecosystem of the surrounding land area will be altered as soon as dam construction begins.

1. Impact of Size and Type of Hydropower Plant

It is difficult to correlate the damage caused by dams to their size or type, as the impacts depend on local conditions. Generally plants with smaller dams are considered less environmentally damaging than those with larger dams. Also, run-of-river (ROR) hydropower plants are generally less damaging than reservoir power plants, because it is not necessary to flood large areas upstream of the project for storage. Yet in some cases run of river impacts can also be severe due to river diversion over long stretches of the river.

2. Impact of River Diversion

While both ROR and reservoir types of hydropower dams may divert water, this is always the case with ROR plants, since they seek to increase kinetic energy with an increased head. The length of diversion can range from a few meters or less to kilometers (km). For example, the Teesta-V ROR dam in north eastern India diverts water for a 23 km long stretch of the river. Eventually the diverted water is returned to the river.

Often downstream flows are reduced considerably or even completely stopped during certain periods of time with sudden intervals of high flows. Such drastic variability in water flow impacts the structure of aquatic ecosystems often leading to a loss of biodiversity. Also, under normal conditions, increased sediment transport from low to intermediate flows provides a warning to aquatic organisms that high flows may follow. Abrupt changes from low to high flows obliterate this cue, making it difficult for organisms to respond to impending environmental changes. A decrease in fish populations has been observed in dewatered reaches below diversions. After long periods of little to no flow some species may not be able to recover and go extinct.

3. Impact of the Reservoir

Dams have major impacts on the physical, chemical and geo-morphological properties of a river. Environmental impacts of dams have largely been negative. Worldwide, at least 400,000 square kilometers have been flooded by reservoirs.

Once the barrier is put in place, the free flow of water stops and water will begin to accumulate behind the dam in the new reservoir. This land may have been used for other things such as agriculture, forestry, and even residences, but it is now unusable. The loss of habitat may not seem severe but if this area was home to a threatened or endangered species, the dam construction could further threaten that species risk of extinction.

Sedimentation

Large dams with reservoirs significantly alter the timing, amount and pattern of river flow. This changes erosion patterns and the quantity and type of sediments transported by the river. Sedimentation rate is primarily related to the ratio of the size of the river to the flux of sediments. The reservoir that has been rapidly filling up with water immediately begins filling up with sediment as well. The trapping of sediments behind the dam is a major problem. Every year it is estimated that 0.5 to 1% of reservoir storage capacity is lost due to sedimentation. The engineering problem with sedimentation is that less power is generated as the reservoir's capacity shrinks.

Downstream Erosion

Trapping of sediments at the dam also has downstream impacts by reducing the flux of sediments downstream which can lead to the gradual loss of soil fertility in floodplain soils. Clean water stripped of its sediment load is now flowing downstream of the dam. This clean water has more force and velocity then water carrying a high sediment load and thus erosion of the riverbed and banks becomes problematic. Since this is unnatural and a form of "forced erosion" it occurs at a much faster rate then natural river process erosion to which the local ecosystem would be able to adapt.

Impact on Local Climate

Another often-ignored environmental effect of the reservoir is the impact on the micro-climate level. Studies indicate that man-made lakes in tropical climates tend to reduce convection and thus limit cloud cover. Temperate regions are also impacted with "steam-fog" in the time period before freezing. Since water cools and warms slower then land, coastal regions tend to be much more moderate then land-locked regions in terms of temperature. Therefore, large dams have a slight moderating effect on the local climate.

Greenhouse Gas Emission from Dams

Freshwater reservoirs can emit substantial amounts of the greenhouse gases methane and carbon dioxide as organic matter submerged in a reservoir decays under anaerobic and aerobic conditions, respectively. Studies indicate that GHG emissions from hydropower reservoirs in boreal and temperate region are low relative to the emissions from fossil fuel power plants, but higher relative to life cycle emissions from wind and solar power.

Tropical reservoirs with high levels of organic matter and shallow reservoirs have higher emission levels. A recent compilation of greenhouse gas emissions from reservoirs found a correlation between the age of the reservoir and latitude. Younger reservoirs and those in low latitudes are the highest emitters. For example, of four Brazilian dams in the Amazon, showed that the GHG emissions factor of the electricity produced by those hydropower dams exceed those from a coal-fired power plant.

Dams Inducing Earthquakes

Finally, a least studied and most disputed physical impact of reservoirs is the possibility of inducing earthquakes. Many scientists believe that seismic activity can be attributed to the creation of dams and their adjacent storage reservoirs. They postulate that the added forces of the dam along inactive faults seem to free much stronger orogenic tensions. Early research indicates that the depth of the water column may be more important to inducing earthquakes rather then total volume of water in the reservoir. While these impacts can be quite severe often they do not receive the attention of the biological impacts that people tend to associate more with animals like fish.

Impact on Fisheries

Dams and river diversion can impact freshwater, as well as marine fisheries. Estuarine and marine fisheries are dependent on estuaries and rivers as spawning grounds and the transport of nutrients from the river to the sea. Migratory fish are especially vulnerable to the impacts of dam construction. Dams can prevent migrating fish such as salmon and eel to reach their spawn grounds.

Production of Biomass Energy

The organic material is converted into usable form known as bio-energy. The materials used in the process of energy production are termed as feedstock.

To better understand biomass, we will explore the various sources first.

Biomass production refers to the increase in the amount of organic matter. It is the addition of organic matter in a given area or population. Biomass is considered renewable energy because it is replenished as plants and animals grow.

There are two forms of production –

Primary production refers to the generation of energy by plants through photosynthesis. The excess energy generated is stored and adds up to the total biomass in the ecosystem. Primary production could be estimated from the total forest cover in a given year.

Secondary production is the absorption of organic matter as body tissues by organisms. It includes ingestion by animals i.e. feeding, whether on other animals or on plants. It also involves decomposition of organic matter by microorganisms. Secondary production could be estimated as the total meat produced per year.

Though biomass could be measured as mass of organisms living and dead in a given environment, production is harder to estimate. It can only be estimated as the increase in volume though part of the additional biomass may have been replaced through natural processes.

Direct Combustion for Heat

Direct combustion for heat is the oldest method of biomass conversion to energy since the earliest civilizations. Thermochemical conversion (combustion) could be achieved in a number of ways using varied feedstock.

Standalone Combustion

Biomass based generators use diesel derived from vegetable oils to fuel diesel generators. The generators burn the organic diesel to produce energy to produce electricity.

- Combined heat and power plants are known to cogenerate electricity and useful heat energy. Ceramic industries utilize the heat in drying products such as clay tiles.
- Some power plants use biomass to heat water and produce steam for electricity generation. The biomass is burnt to produce enough heat to boil water.
- Municipal solid waste plants burn solid wastes to generate electricity. This type is prone to criticism since solid wastes mostly contain toxic gases from plastics and synthetic fibers.

Biomass Co-combustion

Apart from stand-alone combustion, biomass could be blended with other fossil fuels and burnt to generate energy. This is called co-firing.

- Biomass could be directly burnt as coal. This is referred to as direct co-combustion.
- In other cases, the biomass is first processed to gas and then converted to syngas.
- The third case is where fossil fuel is burnt in a different furnace and the energy produced is then used to preheat water in a steam power plant.

Types of Combustion

The various types of combustion are –

- Fixed bed combustion: This is a method where solid biomass is first cut into small pieces and then burnt on a flat fixed surface.
- Moving bed combustion: In this method, a grate is set to constantly and evenly move leaving ash behind. The fuel burns in combustion levels.
- Fluid-bed combustion: Fuel is boiled under high pressure mixed with sand. The sand serves to distribute the heat evenly.
- Burner combustion: In this method, wood dust and fine dust are placed in a burner similar to that of liquid fuel.
- Rotary furnace combustion: A kiln furnace is used to burn organic matter with high moisture content. Such waste as food residue or other moist farm waste is burnt this way.

Pyrolysis

Pyrolysis is another form of processing bio-fuels by burning under very high temperatures without oxygen, which could cause complete combustion. This causes irreversible physical and chemical changes. The absence of oxidation or halogenations processes results in a very dense bio-fuel that could be used in combustion, co-combustion or converted to gas.

- Slow pyrolysis occurs at about 400° C. It is the process of making solid charcoal.
- Fast pyrolysis occurs between 450° C to 600° C and results in organic gas, pyrolysis vapor, and charcoal. The vapor is processed by condensation to liquid form as biooil. This must be

done within 1 second to prevent further reaction. The resultant liquid is dark brown liquid denser than wood biomass and has equal content in terms of energy.

Bio-oil has a number of advantages. It is easier to transport, burn, and store. Many kinds of feed-stock can be processed through pyrolysis to produce bio-oil.

The diagram given below explains the process in converting energy in to a usable form from bio-fuels through Pyrolysis.

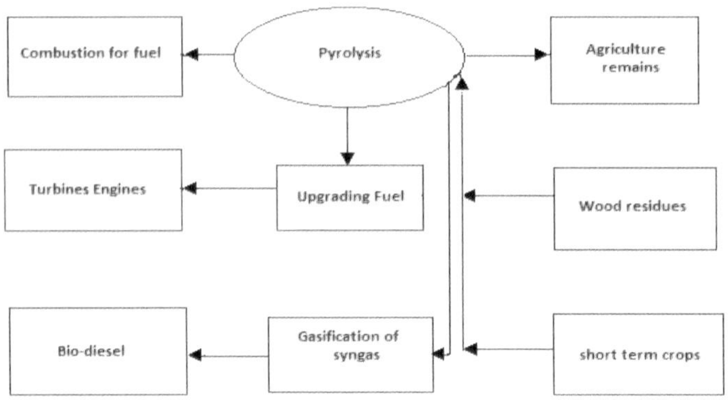

Alcoholic Fermentation

Alcoholic fermentation is the process that converts sugars into cellulose. The process results in ethanol and carbon dioxide as the by-products. This process is considered anaerobic since it takes place in the absence of oxygen. Apart from bread baking and manufacturing alcoholic beverages, this process produces alcoholic fuel. The chemical formula for alcoholic fermentation is given by –

$$C_6H_{12}O_{6(aq)} \rightarrow 2CH_3CH_2OH_{(aq)} + 2CO_{2(g)}$$

Sugarcane is the main feedstock for this process especially in dry environments. Corn or sugar bits are used in temperate areas.

Application of Products

The products have the following applications:

- Acetone is a product used for production of food additives, dissolving glue, thinning of paint, grease removers and in cosmetic products.
- Hydrogen is used as a cooling agent in power industry. It is also used in hydrogen cells for energy production.
- Butanol provides better fuel than ethanol. It is also used as an ingredient in paint, cosmetic products, resins, dyes, polymer extractions and in the manufacture of synthetic fiber.
- Ethanol is used as fuel, paint component, and an additive in antiseptics. It is also used in alcoholic beverages.

Anaerobic Digestion of Biogas

Anaerobic digestion is the biological process by which organic matter is broken down to produce biogas in the absence of oxygen. Microorganisms such as Acidogenetic bacteria and acetogens convert the biodegradable matter to biogas. Apart from being a source of energy, it is also a waste deposition method and environmental conservation technique.

The main equation for this conversion that yields carbon dioxide and methane is as follows:

$$C_6H_{12}O_6 \rightarrow 3CO_2 + 3CH_4$$

The step-by-step process is explained below:

- Step 1: Breakdown of organic matter to sizable molecules for conversion. This process is known as hydrolysis.
- Step 2: Acidogens act on the decomposed matter converting them into volatile fatty acids (VFAs) alongside ammonia, CO2 and hydrogen sulfide. The process is called acidogenesis.
- Step 3: The VFAs are further broken down into acetic acid, carbon dioxide and hydrogen.
- Step 4: The final stage is the combination of emissions above to produce methanol, carbon dioxide, and water.

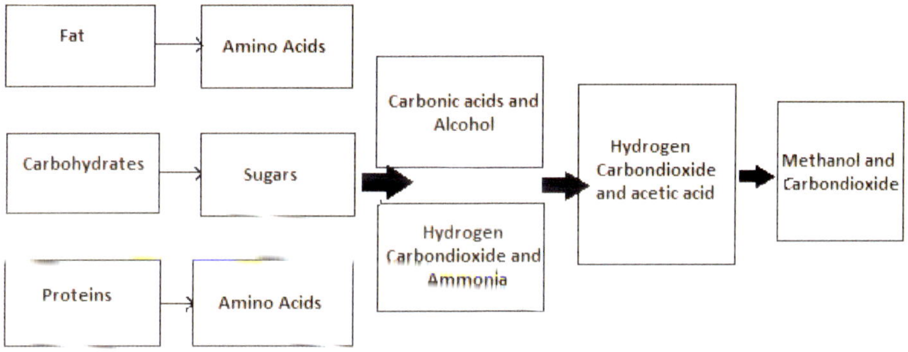

Efficiency of Biomass/Biogas Power Plants

The generation of electricity from biomass results not only in electricity, but also a lot of heat. A traditional gas engine will have an efficiency of 30-35 percent. Gas turbines and steam turbines will end up at around 50 percent.

It is common also use the from these processes, thus increasing the overall energy efficiency. In countries with good governmental support for renewable energy, such as Germany and the Netherlands, there are many plants that combine heat and power production from biomass.

The biogas plant cost does indeed go up, but the long term savings that come with better energy efficiency is in most cases worth it.

As with any other energy source, biomass energy has advantages and disadvantages for its use.

Advantages of Biomass Energy

- Biomass energy can provide a viable solution for what to do with excessive organic waste, such as lumber waste and agricultural waste.
- Because biomass energy production produces both electricity and heat, gas turbines and steam turbines have an approximate 50% efficiency rate, compared to a traditional gas engine with an efficiency of 30-35%.
- The long-term energy savings of biomass energy typically outweigh the expenses that are associated with biogas plant construction and upkeep.
- Biomass energy can be a net carbon neutral energy source if the biomass crops are replanted after harvesting, thus absorbing the amount of CO_2 emissions during their lifecycle that were generated from the burning of biomass.

Disadvantages of Biomass Energy

- Biomass may be needed for soil enrichment and compost instead of burning it for energy production purposes.
- Net greenhouse gases can be produced when biomass feedstock's are burned and they are not sufficiently replanted.
- Biomass can be harvested at unsustainable rates, cause damage to ecosystems, cause air pollution, and require large amounts of water for growth.

References

- How-is-Electric-Energy-Produced-74983: news.softpedia.com, Retrieved 12 March 2018
- Environment-and-health-in-electricity-generation, energy-and-the-environment, information-library: world-nuclear.org, Retrieved 20 July 2018
- How-do-batteries-store-an: scientificamerican.com, Retrieved 19 June 2018
- Non-renewable-energy, encyclopedia: nationalgeographic.org, Retrieved 11 July 2018
- Harmful-effects-of-non-renewable-resources-on-the-environment: greentumble.com, Retrieved 28 March 2018
- The-production-of-electric-power, argomento: eniscuola.net, Retrieved 27 May 2018
- How-oil-is-used-to-make-electricity: oilinformation.weebly.com, Retrieved 17 July 2018
- Electricity-generation-renewable-energy-sources: nationalgeographic.org, Retrieved 16 May 2018

3

Energy Supply

The delivery of fuels for the purpose of consumption is termed as energy supply. This chapter discusses in detail about energy supply, energy requirement, domestic energy consumption, world energy consumption, etc. in the present day scenario.

The quantity of energy supplied is the flow of energy brought onto the market, and the quantity of energy demanded is the amount of energy purchased for a particular period of time. Quantity can be measured in terms of the number of kilowatt hours produced by an electric generator in a day, the number of barrels of oil or cubic feet of gas brought to the market in a month, or the number of tons of coal produced and sold in a year. The responsiveness of fuel supplies to various economic variables are important inputs into energy planning and policy design. For example, if fuel consumption is very responsive to income growth, then high-income growth areas, such as the Asia Pacific regions prior to 1997, will have increased needs for fuel production and imports and the infrastructure that accompanies such fuel growth. Also, if production is very price responsive, only a small increase in price may be necessary to elicit the necessary production to satisfy consumption growth. Such responsiveness on supply side of the market also influences the effects that taxes, subsidies, and a variety of energy policies, have on the price and consumption of energy products.

Competitive Markets

One way, that economists measure such responsiveness of production and consumption, is through supply elasticities. To develop such elasticities, we must first consider supply equations in a competitive market. A competitive market is a stylized market that satisfies a number of properties. First there are so many buyers and sellers that no one of them can influence the price. Second the product sold is homogenous, such as electricity, in which each kilowatt hour sold is like any other one. Third, there is free entry and exit. Thus, if the market is very profitable, firms will enter, but if firms are losing money, they will exit. Fourth in perfect competition, there is perfect information as well. If the above four properties hold, we can represent producers by a supply equation.

Although there are very few perfectly competitive energy markets, many might be workably competitive. Coal is not homogenous but varies by energy, ash, and sulfur, content. Very large firms such as ExxonMobil produce and sell gasoline and may have some pricing power. Nor is all gasoline perceived as being exactly the same. However, again, the various gasolines are very good substitutes and ExxonMobil must compete against other very large firms, such as BP Amoco and TotalFinaElf, which limits its pricing power. Also, firms can and do enter and exit gasoline refining as the market changes. Certainly no market has perfect information, but in capitalist energy markets, there is a large of amount of information published and readily available.

Thus, many energy markets are workably competitive, and supply and demand equations and the resulting elasticities provide a reasonable approximation of how the two sides of the market will behave. Such equations can be estimated and will provide valuable information that can be used for forecasting and policy analysis. In markets where governments intervene and allow or require supply monopoly, supply equations will not be meaningful. This has often been the case for electricity generation, or where very large suppliers have monopoly power and can influence price, such as OPEC in the world oil market. Alternatively, where a large buyer has monopsony power and is able to influence the price, This would be the case for Eskom, the South African electrical utility, which buys all the coal from certain mines. These markets would not be workably competitive, and supply and demand equations and their resulting elasticities would be difficult to develop.

Supply Functions

In a workably competitive market, the quantity of energy supplied to a competitive market is a function of its price along with economic and technological variables that determine costs. For example, in a competitive market the quantity of gasoline, Qs, supplied will be a function of the price received for the gasoline, P, the price paid for the resource inputs such as crude oil and catalysts, Pi, the price of the capital for the refinery necessary to produce the gasoline, P_k, the price of labor, which includes wages, salaries, and indirect labor costs, such as employment taxes, Pl, the price of using land or any other natural resource or other factor of production, P_n, and technological change that improves the production process for this fuel, T. Although the supplier responds to the market prices that they have to pay, if there are market externalities, such as environmental damage, that are not reflected in the market price. Then the market will underprice the good and too much of the good will be produced.

Prices of other related goods will also influence quantity supplied. For example, when heavier fuels are cracked, both gasoline and lighter products are formed. These lighter products produced with gasoline are called complementary goods in production. Increasing their price, Pc, will make gasoline production more attractive. Alternatively, goods may be substitutes in production. A refiner may crack heavier products and produce more gasoline when these heavier substitute product prices, Ps, increase.

Governments may interfere in energy markets and their policies may influence quantity supplied. For example, environmental regulations, E_r, that require less pollution or more safety when producing fuels, will decrease the quantity of fuel supplied. Such regulations include the removal of sulfur from fuels, restrictions on the use of lead in gasoline in many countries in the world, and health and safety requirements in the workplace. Increasing environmental regulation increases production cost and should decrease quantity supplied. The market supply also depends on the number of suppliers in the industry.

Using the above variables, we can write a supply function for any energy source in a competitive market as

$$Q_s = f(\overset{+}{P}, \overset{-}{P_i}, \overset{+}{P_c}, \overset{-}{P_s}, \overset{-}{P_l}, \overset{-}{P_k}, \overset{-}{P_n}, \overset{+}{T}, \overset{-}{E_r}, \#\overset{+}{S})$$

Where P = the own price, P_i = the price of resource inputs, P_c = the price of complementary goods, P_s = the price of substitutes, P_l = the price of labor, P_k = the price of capital, P_n = the price of land or other natural resources, T = technical change, E_r = environmental regulation and policy, and #S = number of suppliers.

The signs above the variables indicate whether the variable is inversely (–) or directly related (+) to gasoline production. If inversely related, the variables move in the opposite direction. When the variable goes up, production goes down. If directly related, the variable and production both increase or both decrease together. Thus, increasing costs of production, the price of substitute goods, and environmental regulations (P_i, P_l, P_k, P_n, P_s, E_r) will all decrease production, while increasing the price of complements in production, improvement in technology, and increasing the number of suppliers, (P, P_c, T, #S) will all increase production.

Supply Elasticities

The responsiveness of quantity supplied to a variable is called the elasticity of supply with respect to that variable. It is the percentage change in quantity divided by the percentage change in the variable. We can write the elasticity of supply with respect to price P as:

$$\varepsilon_s = \frac{\%\text{change}Q_s}{\%\text{change}P} = \frac{\Delta Q_s / Q_s}{\Delta P / P}$$

where Δ represents a discrete change in the variable. If the supply price elasticity of coal is 0.89, then when the price of coal increases by 1%, the quantity of coal supplied increases by 0.89%. The cross-elasticity of supply indicates how quantity supplied is related to price of another good. For example, if the cross-elasticity of gasoline supply with respect to the price of distillate is –0.2, then if the price of distillate increases by 1%, the quantity of gasoline produced decreases by 0.2%.

The time period over which we measure the supply elasticity influences the size of the elasticity. In the short run, say one year, if the price of coal goes up, coal mines may only be able to increase production by a small amount. Since coal mining is very capital intensive, using specialized equipment, it takes time to buy new equipment and typically takes 4-7 years to open a new mine. Thus, the short run elasticity may be quite low. However, in the long run, or in the amount of time required to totally adjust to a price change, production may change much more and the long run elasticity is likely to be much larger than the short run elasticity. The more capital intensive the industry and the longer-lived the capital stock, the greater the difference between the long and the short run elasticities.

Although the long run may vary from product to product, it is fairly well defined. It is simply the time it takes for total adjustment to take place. The short run is less well defined and typically depends on the period of interest, and is most often a year or less. If carrying out statistical analysis using actual data, the short run is often the periodicity of the data. (For example, periodicities include daily or weekly energy prices, monthly data for natural gas in storage, quarterly data on economic indicators, or annual data for oil consumption). If the analysis were on monthly data for coal supply, the short run would be a month. If the analysis were on annual data, the short run would be a year and one would expect the annual response to be larger than the monthly response. The long run elasticities in both cases should be the same but would be more elastic. For example, it has been found that the underground supply of US coal has a short run annual elasticity of 0.61 and a long run elasticity of 1.31. Thus, with time to adapt, coal producers are over twice as price responsive as they would be in a year.

Using Elasticities to Forecast Supply

Supply elasticities are quite useful for policy and planning as shown in the following examples.

Suppose the world price of coal goes up. For example, let the coal price go from US$20 per metric tonne to US$22 per metric tonne. Suppose that the short run (one year) supply elasticity is 0.10 and the long run supply elasticity is 1.1. Since

$$\varepsilon_s = \frac{\Delta Q_s / Q_s}{\Delta P / P}$$

then in the short run

$$\Delta Q_s / Q_s = \varepsilon_s (\Delta P / P) = 0.10 \times 2 / 20 = 0.01$$

Thus, Australian production would go up 0.01 or 1 percent. Australian production was 345.664 million metric tons in 1998, thus production would increase 0.01×345.664 = 3.456 million metric tonnes.

In the long run, the increase would be

$$\Delta Q_s / Q_s = \varepsilon_s (\Delta P / P) = 1.1 \times 2 / 20 = 11\%$$

for an increase of 0.11×345.664 = 38.023 million metric tonnes.

The same computations could be done for any policy that changed right hand side variables in the supply equation for any market that is considered competitive. For example, Israel has the largest density of solar hot water heaters supplying around 80 percent of the Israeli market for hot water. If interest rates lowered capital costs for Israeli solar hot water equipment by 22 percent, we can compute what would happen to the short run production of such hot water heaters. Suppose the short run cross price elasticity of supply with respect to the price of capital is −0.15. We can again use the same procedure but substituting the proportionate change in capital price for the proportionate change in own price or:

$$\Delta Q_s / Q_s = \varepsilon_s (\Delta P_k / P_k) = (0.15) \times (-0.22) = 0.033$$

If production of water heaters had been 100,000, then new production would be (1 + 0.033) ×100,000 = 103,300.

Supply elasticities for products are derived by using econometric estimations based on historical data, or by using engineering estimates of costs of production. Studies using simple models for the US have found the following long run own price elasticities of supply: ε (natural gas) = 0.41, ε (uranium) = 0.74 to 3.08, ε (Appalachian coal) = 0.41 to 7.90.

Total Primary Energy Supply

Total primary energy supply (TPES) is the total amount of primary energy that a country has at their disposal. This includes imported energy, exported energy (subtracted off) and energy extracted from natural resources (energy production). Usually TPES is thought of as being the sum of all primary energy sources, but some end use energymust be taken into consideration. This comes from the fact that TPES includes both imports and exports. The net amount of electricity traded (imports - exports) and the net secondary fuels (for example amount of gasoline imported - amount of gasoline exported) become part of the TPES.

In contrast to the total primary energy supply, countries have a total final consumption of energy, which focuses on the end use energy.

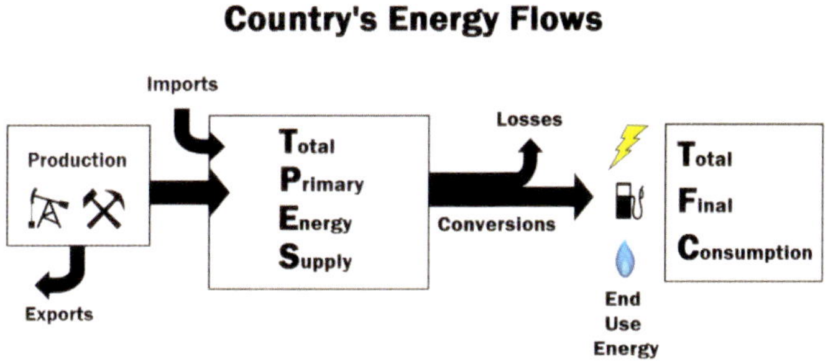

Figure: This diagram shows how Total Primary Energy Supply becomes Total Final Consumption. Various primary energy sources combine are changed with energy conversion technologies like power plants and refineries to energy currencies.

Energy Policy

Policy is the response of government to the prevailing circumstances in a particular sphere of activity in a society. Government has a choice either to maintain the prevailing circumstances because they need to meet certain social objectives, or to change those circumstances so as to ensure that certain social objectives will be met. Thus policy may consist of positive actions by government or simply the maintenance of the status quo.

Energy policy is also recognition that economic factors alone do not determine the outcome in the energy sector. Governments take on a responsibility for ensuring that the outcomes in the energy sector are consistent with a range of social objectives. If the only objective was economic efficiency in the energy sector then it could be argued that the only role for government is the maintenance of competition or the regulation of natural monopoly.

The energy sector however does not sit isolated from other spheres of activity in a modern society. In addition, within the energy sector itself competition is not sufficient even to produce efficiency outcomes where problems of externalities are involved.

Thus, energy policy has to be concerned with not only efficiency but also concerns associated with social equity and the issues associated with environmental impacts of energy use, as well as long-term sustainability of the energy resources available to society. In addition, energy policy has to be made consistent with policy in other policy areas such as transportation, the built environment, industry policy, and taxation policies.

Increasingly energy policy conducted by national governments has to be operated in an international context. The prevailing perspective is one of the "global common," in which the energy policy actions of one country spill over in to positive or negative impacts on other countries. Thus national governments are not free to conduct energy policy in isolation from each other. The greenhouse issue in particular has given rise to a requirement for coordinated energy policies between countries so as to achieve not just national policy objectives, but also international policy objectives.

A range of "instruments" is available to governments to enable them to achieve their policy objectives. However, even if energy policy objectives are well defined, the instruments for achieving such objectives may not be so well defined. In addition, as the value of policy objectives changes continuous evaluation is required in order to ensure the effectiveness of the policy instrument "mix" available to the government. These evaluation processes raise the question as to which groups participate in the policymaking processes and the role of consultation by government over energy policy issues.

Energy policy can be considered on a number of levels. At the deepest level are those fundamental values that seek to address questions such as: what sources of energy should be used; who should own energy resources; and how should energy resources be utilized. This can include issues such as whether the government should promote a highly integrated energy supply system or a more diffused energy supply system. In the latter these fundamental value questions may be addressed by a range of individuals and groups according to their own preferences rather than through a centralized energy supply authority.

At another level there is scope for a genuine public debate over the choice of energy policy instruments to achieve objectives for which there may be general agreement. Examples may include debate of the merits of nuclear energy as a solution to the greenhouse problem, or subsidies for renewable energy technologies with the same intention.

Energy policy, therefore, is not just a matter of economics. It traverses the entire spectrum of concerns in society and highlights the role of government as the body established by society and empowered to deal with these concerns.

National Energy Policy

Measures used to Produce an Energy Policy

A national energy policy comprises a set of measures involving that country's laws, treaties and agency directives. The energy policy of a sovereign nation may include one or more of the following measures:

- Statement of national policy regarding energy planning, energy generation, transmission and usage
- Legislation on commercial energy activities (trading, transport, storage, etc.)
- Legislation affecting energy use, such as efficiency standards, emission standards
- Instructions for state-owned energy sector assets and organizations
- Active participation in, co-ordination of and incentives for mineral fuels exploration and other energy-related research and development policy command
- Fiscal policies related to energy products and services (taxes, exemptions, subsidies)
- Energy security and international policy measures such as:
    - International energy sector treaties and alliances,

- o General international trade agreements,
- o Special relations with energy-rich countries, including military presence and/or domination.

Frequently the dominant issue of energy policy is the risk of supply-demand mismatch. Current energy policies also address environmental issues, particularly challenging because of the need to reconcile global objectives and international rules with domestic needs and laws. Some governments state explicit energy policy, but, declared or not, each government practices some type of energy policy. Economic and energy modelling can be used by governmental or inter-governmental bodies as an advisory and analysis tool.

Factors within an Energy Policy

There are a number of elements that are naturally contained in a national energy policy, regardless of which of the above measures was used to arrive at the resultant policy. The chief elements intrinsic to an energy policy are:

- What is the extent of energy self-sufficiency for this nation
- Where future energy sources will derive
- How future energy will be consumed (e.g. among sectors)
- What fraction of the population will be acceptable to endure energy poverty
- What are the goals for future energy intensity, ratio of energy consumed to GDP
- What is the reliability standard for distribution reliability
- What environmental externalities are acceptable and are forecast
- What form of "portable energy" is forecast (e.g. sources of fuel for motor vehicles)
- How will energy efficient hardware (e.g. hybrid vehicles, household appliances) be encouraged
- How can the national policy drive province, state and municipal functions
- What specific mechanisms (e.g. taxes, incentives, manufacturing standards) are in place to implement the total policy
- What future consequences there will be for national security and foreign policy

State, Province or Municipal Policy

Even within a state it is proper to talk about energy policies in plural. Influential entities, such as municipal or regional governments and energy industries, will each exercise policy. Policy measures available to these entities are lesser in sovereignty, but may be equally important to national measures. In fact, there are certain activities vital to energy policy which realistically cannot be administered at the national level, such as monitoring energy conservation practices in the process of building construction, which is normally controlled by state-regional and municipal building codes (although can appear basic federal legislation).

Americas

Brazil

Brazil is the 10th largest energy consumer in the world and the largest in South America. At the same time, it is an important oil and gas producer in the region and the world's second largest ethanol fuel producer. The governmental agencies responsible for energy policy are the Ministry of Mines and Energy (MME), the National Council for Energy Policy (CNPE, in the Portuguese-language acronym), the National Agency of Petroleum, Natural Gas and Biofuels (ANP) and the National Agency of Electricity (ANEEL). State-owned companies Petrobras and Eletrobrás are the major players in Brazil's energy sector.

United States

Currently, the major issues in energy policy revolve around the rapidly growing production of domestic and other North American energy resources. The drive toward energy independence and less reliance on oil and coal is fraught with partisan conflict because these issues revolve around how best to balance both competing values, such as environmental protection and economic growth, and the demands of rival organized interests, such as those of the fossil fuel industry and of the newer renewable energy businesses.

Europe

European Union

Although the European Union has legislated, set targets, and negotiated internationally in the area of energy policy for many years, and evolved out of the European Coal and Steel Community, the concept of introducing a mandatory common European Union energy policy was only approved at the meeting of the European Council on October 27, 2005 in London. Following this the first policy proposals, *Energy for a Changing World*, were published by the European Commission, on January 10, 2007. The most well known energy policy objectives in the EU are 20/20/20 objectives, binding for all EU Member States. The EU is planning to increase the share of renewable energy in its final energy use to 20%, reduce greenhouse gases by 20% and increase energy efficiency by 20%.

Germany

In September 2010, the German government adopted a set of ambitious goals to transform their national energy system and to reduce national greenhouse gas emissions by 80 to 95% by 2050 (relative to 1990). This transformation become known as the *Energiewende*. Subsequently, the government decided to the phase-out the nation's fleet of nuclear reactors, to be complete by 2022. As of 2014, the country is making steady progress on this transition.

United Kingdom

The energy policy of the United Kingdom has achieved success in reducing energy intensity (but still really high), reducing energy poverty, and maintaining energy supply reliability to date.

The United Kingdom has an ambitious goal to reduce carbon dioxide emissions for future years, but it is unclear whether the programs in place are sufficient to achieve this objective (the way to be so efficient as France is still hard). Regarding energy self sufficiency, the United Kingdom policy does not address this issue, other than to concede historic energy self sufficiency is currently ceasing to exist (due to the decline of the North Sea oil production). With regard to transport, the United Kingdom historically has a good policy record encouraging public transport links with cities, despite encountering problems with high speed trains, which have the potential to reduce dramatically domestic and short-haul European flights. The policy does not, however, significantly encourage hybrid vehicle use or ethanol fuel use, options which represent viable short term means to moderate rising transport fuel consumption. Regarding renewable energy, the United Kingdom has goals for wind and tidal energy. The White Paper on Energy, 2007, set the target that 20% of the UK's energy must come from renewable sources by 2020.

The Soviet Union and Russia

The Soviet Union was the largest energy provider in the world until the late 1980s. Russia, one of the world's energy superpowers, is rich in natural energy resources, the world's leading net energy exporter, and a major supplier to the European Union. The main document defining the energy policy of Russia is the Energy Strategy, which initially set out policy for the period up to 2020, later was reviewed, amended and prolonged up to 2030. While Russia has also signed and ratified the Kyoto Protocol. Numerous scholars note that Russia uses its energy exports as a foreign policy instrument towards other countries.

Asia

India

The energy policy of India is characterized by trades between four major drivers:

- Rapidly growing economy, with a need for dependable and reliable supply of electricity, gas, and petroleum products;
- Increasing household incomes, with a need for affordable and adequate supply of electricity, and clean cooking fuels;
- Limited domestic reserves of fossil fuels, and the need to import a vast fraction of the gas, crude oil, and petroleum product requirements, and recently the need to import coal as well; and
- Indoor, urban and regional environmental impacts, necessitating the need for the adoption of cleaner fuels and cleaner technologies.

In recent years, these challenges have led to a major set of continuing reforms, restructuring and a focus on energy conservation.

Thailand

The energy policy of Thailand is characterized by 1) increasing energy consumption efficiency, 2) increasing domestic energy production, 3) increasing the private sector's role in the energy sector,

4) increasing the role of market mechanisms in setting energy prices. These policies have been consistent since the 1990s, despite various changes in governments. The pace and form of industry liberalization and privatization has been highly controversial.

Bangladesh

The first National Energy Policy (NEP) of Bangladesh was formulated in 1996 by the Ministry of Power, Energy and Mineral resources to ensure proper exploration, production, distribution and rational use of energy resources to meet the growing energy demands of different zones, consuming sectors and consumers groups on a sustainable basis. With rapid change of global as well as domestic situation, the policy was updated in 2004. The updated policy included additional objectives namely to ensure environmentally sound sustainable energy development programmes causing minimum damage to environment, to encourage public and private sector participation in the development and management of energy sector and to bring the entire country under electrification by the year 2020.

Oceania

Australia

Australia's energy policy features a combination of coal power stations and hydro electricity plants. The Australian government has decided not to build nuclear power plants, although it is one of the world's largest producers of uranium.

Energy Consumption

Energy consumption is simply the use of power. Whether you're driving to work, turning on a lamp or charging your computer, you're consuming energy. The more energy you use, the higher your level of energy consumption. Let's look at a few examples:

- In a factory, total energy consumption can be measured by looking at how much energy a production process consumes, for example, by making car parts. This will include water, electricity, gas any energy source needed for the stuffed animals to be made.
- In a household, energy consumption will include the electricity, gas, water, and any other energy used to live in it.
- The energy consumption of a bus includes how much diesel or gasoline it uses to run.

All this to clear up the following: energy consumption does not necessarily come from a single energy source. It is a common misconception to think that to save energy you have to save electricity. It could very well be a totally different energy source that has the greatest impact on a certain process.

Looking closely at the definition of energy consumption and going back to our first example, the car parts factory, means we have to consider how an entire production process is evaluated. If you are part of a value-added process such as industrial metallurgy (like TACSA, for example) the process does not end when the car parts are manufactured. If your delivery fleet takes them to the customer, the gasoline from that fleet also forms part of your energy consumption.

Cost of Energy Consumption Cost

Given the above information, figuring out the answer to this question is easy. In economic terms, your energy consumption is the sum total of your energy bills.

Calculating energy consumption helps to come up with two very interesting figures if you plan to save energy:

1. Knowing how much you would have to invest in an energy management system that would help you achieve continuous, verified savings.
2. Know the savings potential of the building you live or work in, and understand where to begin saving.

Factors which Influence Energy Consumption

Such differences may largely be attributed to a multitude of factors, several of which are listed below:

- plant capacity;

- rate of plant utilization;

- age;

equipment and the manufacturing process adopted;

- integrated or non-integrated operation;

- site layout;

- overall efficiency of operation;

- degree of mechanization of materials handling;

- extent of process automation;

- wood species, size and moisture content;

- product type and mix;

- degree of product finishing;

- proportion of air to kiln drying;

- climatic conditions;

- energy prices;

- load factor;

- steam and power balances;

- extent of energy conservation measures adopted;

- standard of housekeeping and maintenance.

Energy Consumed by the Industry

Compared to other industries the wood and wood products industry is not considered energy intensive, and, as shown in table below, it consumed some 7.34 mtoe of energy in the OECD countries during the year 1983, accounting for approximately 0.3 percent of the OECD countries primary energy demand.

Table: Energy Consumption in the OECD Countries

	Electricity (mtoe)	Oil (mtoe)	Gas (mtoe)	Solid Fuels (mtoe)	Total (mtoe)	%
TOTAL CONSUMPTION	395	1 362	491	280	2 528	100
inclusive of						
TOTAL INDUSTRY	166	246	202	208	822	33
including:						
Iron and steel	20	9	20	92	141	6
Chemicals:	34	131	57	14	236	9
Feedstocks	-	112	10	-	122	5
Food & tobacco	10	15	16	6	47	2
Pulp, paper & printing[1/]	18	15	14	12	59	2
Wood & wood products[1/]	3	2	1	1	7	0.3

Does not include energy derived from process residues

However, on a national scale the proportion of energy consumed in the wood and wood products industry in relationship to industry as a whole, will vary according to the degree of its importance to the national economy, as indicated in table. In the Nordic countries where the first industry, including pulp and paper, is regarded as one of the major manufacturers in their industrial sector, it accounts for 15 percent of energy input to the manufacturing industry, and in the case of Finland their forest industry represents some 46 percent.

Table: Purchased energy consumption by the wood and wood products industry of selected OECD countries (1983)

Country	Purchased energy consumption Total industry (mtoe)	Wood & wood products industry (mtoe)	(%)
Austria	5.98	0.12	2.0
Canada	50.58	0.66	1.3
Finland	9.16	0.47	5.1
France	46.10	0.32	0.7

Germany W.	67.14	0.60	0.9
Norway	7.01	0.22	3.1
Portugal	3.85	0.10	2.6
Spain	19.82	0.18	0.9
Sweden	11.90	0.67	5.6
USA	338.45	2.98	0.9
OECD Total	822	7.35	

Reduce Consumption, Increase Conservation

Many institutions are working to create a healthier planet by reducing energy consumption and increasing resource conservation.

- Sports – From efficient field irrigation to game day recycling programs and stadiums built with environmental design in mind, many sports teams have a game plan for saving energy this season.
- Schools – Universities across the are implementing plans to become more energy efficient. At Clemson University, students and faculty use leftover cooking oil from on-campus dining halls to create renewable biodiesel, which is used to fuel the school's facilities trucks. Curious to see how Clemson's rivals stack up on sustainability? Check out our rankings for the greenest top football schools.
- Social – Social sharing sites like Pinterest offer thousands of creative and innovative ideas for turning trash into recycled treasure. Recycling helps reduce consumption of natural materials, eliminate energy use during manufacturing and conserve the gasoline used to transport new products to store shelves. We can get you started with some upcycling ideas for glass jars, plastic bottles and just about anything else you have lying around.

World Energy Consumption

Global energy consumption has about doubled in the last three decades of the past century. In 2004, about 77.8% of the primary energy consumption is from fossile fuels (32.8% oil, 21.1% natural gas, 24.1% coal), 5.4% from nuclear fuels, 16.5% from renewable resources, of which the main one is hydroelectric, 5.5%, whereas the remaining 11% consists of non commercial biomasses, such as wood, hay, and other types of fodder, that in rural-economies still constitute the main resource.

These 'rural' biomasses (mainly fodder) are usually ignored by statistical reviews of energy consumption proposed by oil companies, but for a correct global perspective they ought to be considered, because at least two thirds of human kind still lives in rural and artisanship economies not too different from the European Middle Age. For example, in the USA in 1850 two thirds of the mechanical energy was obtained from horses, and in 1925 horses were still 30 million.

To fix ideas, the amount of energy currently consumed everyday by human kind corresponds to the heating value of 100 large oil tankers. Not accounting for the 1.2 Gtoe/yr of 'rural biomasses',1 the fraction of energy demand covered by fossil fuels in 2004 appears to be 87.7%, a percentage often cited by various sources. Direct solar energy usage is about 11 Mtoe (millions, not billions of toe), less than 0.1% of the global consumption.

Before broadening the temporal perspective, we discuss current consumption vs. population in different regions of our planet. Main data are taken from

In table below nations are grouped by type of economy, level of industrial development and energy per-capita consumption intensity. Large differences between groups of nations are due to several factors.

Table: World energy consumption, population and energy intensity in year 2000 by homogeneous groups of nations with similar type of economy and levels of industrial development and energy intensity

Consumption year 2000	Consumption	Population	Consumption per capita	
homogeneous group of countries	(Mtoe/yr)	(billions inhabs)	toe/yr	kgoe/day
North America	2.4	0.3	7.9	21.7
Japan, Australia, New Zealand	0.6	0.2	3.9	10.6
Europe OECD	1.6	0.5	3.5	9.6
Ex URSS and Central-East Europe	1.3	0.4	3.1	8.4
Latin America	0.8	0.5	1.5	4.1
Middle East	0.5	0.4	1.3	3.6
East Asia	0.6	0.5	1.1	3.1
China	1.4	1.4	1.0	2.6
Africa	0.4	0.7	0.6	1.6
South Asia	0.6	1.4	0.4	1.2
Total globe	10.3	6.2	1.7	0.6

Geoclimatic Factors

Among such factors are, for example, diverse geographic and climatic conditions. A rigid clima requires intense residential heating; low population density implies transportation needs over long distances; high humidity levels in hot seasons call for intense building and vehicle air conditioning.

Another factor is technical-economical-organizational efficiency in energy resource management that depends on the type of internal political-economical system.

For example, the bureaucratic mentality that dominates or has dominated in collectivized economies such as ex URSS and China, where everything is subjected to meticulous central planning, is a cause of dissipation, even when technological level and social liveliness are high.

However, the most important factor that determines per capita consumption is the level of social development and industrialization. Table below shows the historical trend. In the last two millennia, global consumption experienced a 70-fold increase, population a 20-fold increase, and per capita consumption little more than a 3-fold increase (from 0.5 toe/yr to 1.7 toe/yr). The transition from rural biomasses to fossil fuels has accompanied and allowed development and industrialization in selected regions of the globe and has determined profound changes in the quality of human life.

Table: Historical evolution of per capita energy consumption

History of annual per capita energy consumption	toe/yr
Food for survival (3000 kcal/day)	0.11
After discovery of fire (500,000 years ago)	0.22
Neolithic age, bronze age, iron age	0.45
Greek-roman rural-artisan middle-age economy	0.50
1800 – England	0.55
1900 – England	2.8
2000 – England	3.5

Agricultural fraction of gross national product (%)

<1900 – Italy	66	0.50
1900 – Italy	50	0.50
1913 – Italy	42	0.55
1939 – Italy	28	1.0
1981 – Italy	6.40	2.5
2000 – Italy	3.30	3.0

	Global consumption average Gtep/anno	Global population in billions inhabitants	Per capita consumption
Greek-roman age	0.15	300	0.5
Year 2000	10.3	6200	1.7

Technological Factors

In each historical period, increases in energy demand have always gone together with improvement in energy efficiency. Technological and scientific advances have determined and are continuing to determine impressive and regular improvements in the efficiency of energy conversion devices.

It is a process that follows closely the typical logistic law of human learning, as shown by figure below where we graph the regular growth of the thermodynamic efficiency of energy conversion equipment during the last 300 years. The first steam engines for water extraction from coal mines at the dawn of the industrial revolution were <1% efficient. Today's combined cycle and fuel cell power plants reach 55–60% net efficiencies. The logistic feature of this regular improvement, shown by the linear growth of the logarithm of $\eta_{II}/(1 - \eta_{II})$ where η_{II} is the so-called second-law thermodynamic efficiency, indicates with some confidence that scientific learning and technological progress will continue, and by the end of the century will bring second-law thermodynamic efficiency of energy conversion devices to about 80%.

From the analysis of past consumptions, we observe the rapid increase occurred during the last century and can compute the overall past consumption of fossil fuels. In order to estimate the consumption in the current century, we must first estimate population growth.

Regular growth of the second-law thermodynamic efficiency (η_{II}, right) of energy conversion equipment during the last 300 years. The linear 'logistic' growth of the logarithm of $\eta_{II}/(1 - \eta_{II})$ is typical of many learning processes and takes into account the typical saturation of all learning curves. The datum for fuel cells is currently hypothetical

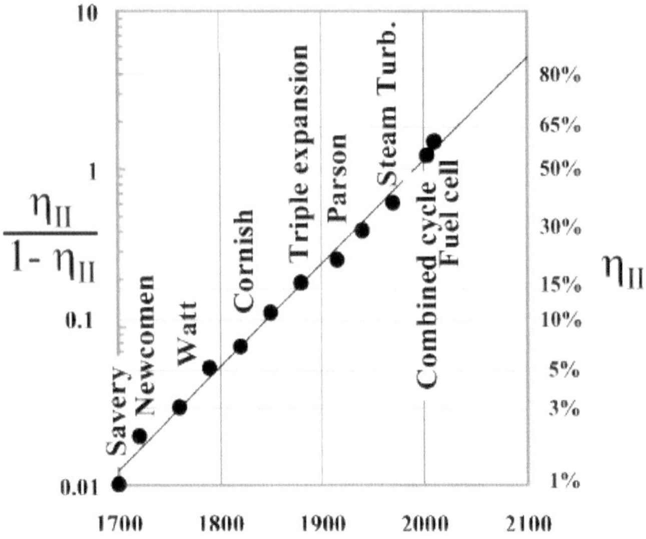

Demographic Factors

Clearly, there is no room on our planet for an indefinite population growth. All credited studies agree with the estimate that a sustainable future requires that population stabilizes at about twice the current population and that this will be achieved around the end of this century. But demographic growth rates will vary from region to region of the globe. To proceed, we adopt estimates based on the (certainly questionable) assumption that the decrease in birth rates experienced in developed countries can be extrapolated also to countries with entirely different cultures.

The graphs in figure above highlight the strong inverse correlation that exists between per capita energy consumption and key factors for demographic growth, such as fertility, infant mortality and life expectancy. Indeed, the use of energy allows improvements in living standards, access to medical care, use of contraception, a longer life expectancy, social services that diffuse alphabetization and access to information, work opportunities for women, lesser need to have a numerous family with enough work power to sustain the unproductive members (children and elders). Per capita energy consumption is at the same time an indicator and an instrument of socioeconomic development.

Nations with high living standards have little or no demographic growth. Underdeveloped nations have high growth rates, in some instances doubling the population in 25 years. An important threshold appears to be that of one toe/yr per capita. Social conditions improve, life expectancy reaches 70 years, fertility and demographics growth slowdown.

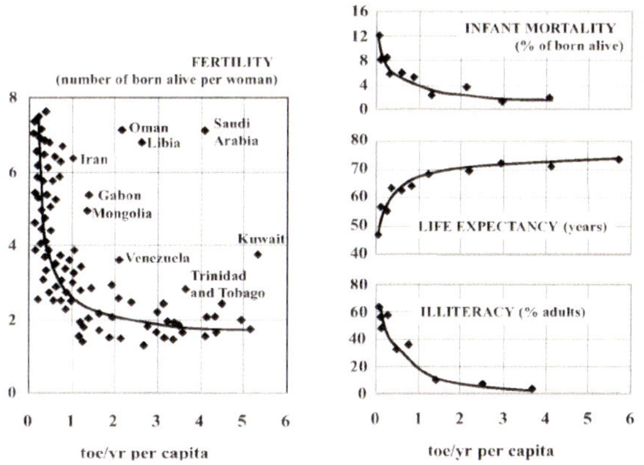

Figure: Correlations between different socioeconomic level indicators and per capita energy consumption

Domestic Energy Consumption

Energy use in buildings is one of the largest contributors to energy consumption both locally and globally. Think about how you use energy every day. You wake up to an alarm clock. You take a shower with water warmed by a hot water heater. You listen to music on the radio as you dress. You catch the bus to school. That's just the energy you use before you get to school! Energy use is sometimes called energy consumption.

Users of Energy

The Department of Energy divides energy users into different categories: residential, commercial, industrial, electric power, and transportation. These are called the sectors of the economy.

Residential and Commercial Sectors

Any place where people live is considered a residential building. Commercial buildings include offices, stores, hospitals, restaurants, and schools. Residential and commercial buildings are often grouped together because they use energy in the same ways—for heating and cooling, lighting, heating water, and operating appliances.

Together, homes and buildings consume almost 11 percent of the energy used in the United States today. In the last 30 years, Americans have reduced the amount of energy used in their homes and commercial buildings. We still heat and cool rooms, and heat hot water. We have more home and office machines than ever. Most of the energy savings have come from improvements in technology and in the ways the equipment is manufactured.

Heating and Cooling

It takes a lot of energy to heat rooms in winter and cool them in summer. Forty-eight percent of the energy used in the average home is for heating and cooling rooms. The three fuels used most often for heating are natural gas, electricity, and heating oil. Today, more than half the nation's homes use natural gas for heating.

Most natural gas furnaces in the 1970s and 1980s were about 60 percent efficient. That means they converted 60 percent of the energy in the natural gas into usable heat. New gas furnaces are designed to be up to 98 percent efficient.

The second leading fuel for home heating is electricity. Electricity also provides almost all of the energy used for air conditioning. The efficiency of heat pumps and air conditioners has increased more than 50 percent in the last 35 years.

Heating oil is the third leading fuel used for home heating. New oil furnaces burn oil more cleanly and operate more efficiently.

In the future, we may see more use of renewable energy sources, such as geothermal and solar energy, to heat and cool our homes and workspaces.

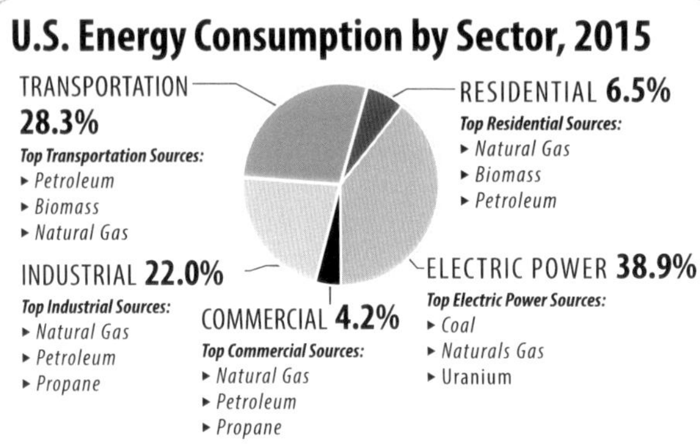

The residential, commercial, and industrial sectors use electricity. This graph depicts their energy source consumption outside of electricity.

Data: Energy Information Administration

*Total does not equal 100% due to independent rounding.

Lighting

Homes and commercial buildings also use energy for lighting. The average home spends five percent of its utility costs on lighting. Schools, stores, and businesses use between 9 and 10 percent of their utility costs for lighting. Most commercial buildings use fluorescent lighting. It costs more to install, but it uses a lot less energy to produce the same amount of light.

Many homes still use the type of light bulb invented by Thomas Edison over 100 years ago. These incandescent bulbs are not very efficient. Only about 10 percent of the electricity they consume is converted into light. The other 90 percent is converted to heat.

Due to the Energy Independence and Security Act of 2007, most traditional, inefficient incandescent light bulbs have been phased out and are no longer for sale. Consumers can choose from several more efficient bulb options as replacements. Energy-saving incandescent, or halogen, bulbs are more expensive than traditional incandescent, but use 25 percent less energy and last three times as long.

Compact fluorescent light bulbs (CFLs) can be used in light fixtures throughout homes. CFLs cost less to use because they last longer and use less energy than incandescent bulbs.

Even more efficient than CFLs, light emitting diodes (LEDs) are available. LEDs have become a very affordable option for lighting in homes and businesses, and use even less energy than CFLs. LEDs last 25 times longer than a traditional incandescent, are more durable than the other options available on the market, and have many technical applications.

Appliances

Over the last 100 years, appliances have changed the way we spend our time at home. Chores that used to take hours can now be done in minutes by using electricity instead of human energy. In 1990, Congress passed the National Appliance Energy Conservation Act, which requires appliances to meet strict energy efficiency standards. As a result of this Act, home appliances have become more energy efficient. Water heaters, refrigerators, clothes washers, and dryers all use much less energy today than they did 25 years ago.

Appliance Efficiency Ratings

When you buy an appliance, you should pay attention to the yellow Energy Guide label on every appliance. This label tells you the Energy Efficiency Ratio (EER) of the appliance. The EER tells how much it costs to operate the appliance.

Payback Period

Whether you buy a furnace, hot water heater, refrigerator, or other home appliance, you must choose the best bargain. Since most high-efficiency systems and appliances cost more than less efficient ones, you have to know how much it will cost to operate the appliance each year and how many years you can expect to use it. The payback period is the amount of time you must use a system or appliance before you begin to benefit from energy savings.

For example, if you buy an efficient refrigerator that costs $100 more, but uses $20 less electricity each year, you would begin saving money after five years. Your payback period would be five years. Since refrigerators usually last ten years, you would save $100 over the life of the appliance and save natural resources.

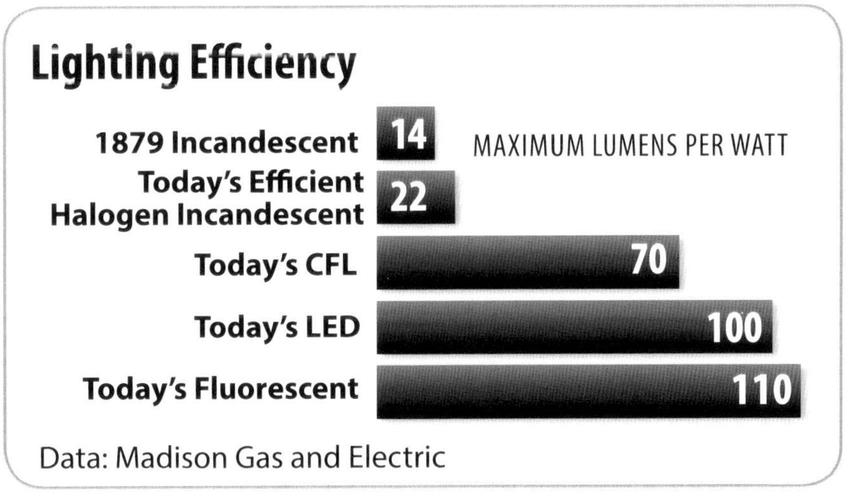

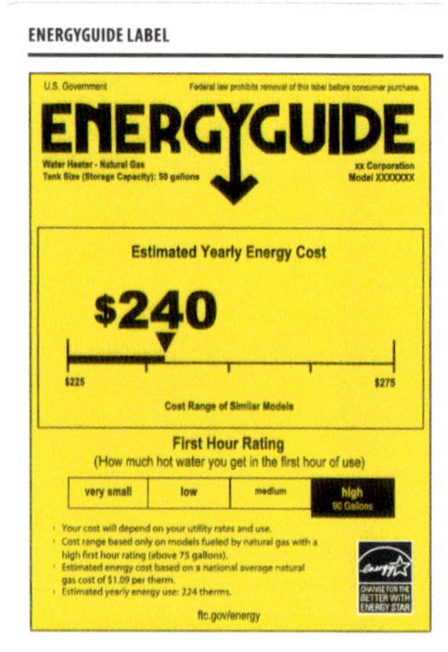

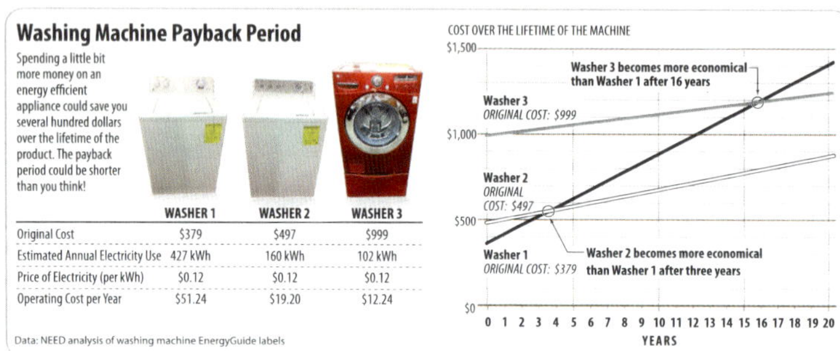

Industrial Sector

Every industry uses energy, but six energy-intensive industries use most of the energy consumed by the industrial sector. However, advanced technologies allow industries to do more with less energy

Petroleum Refining

The United States uses more petroleum than any other energy source. Petroleum provides the with about 38.8 percent of the energy we use each year. Petroleum can't be used as it comes out of the ground. It must be refined before it can be used.

Oil refineries use a lot of energy to convert crude oil into gasoline, diesel fuel, aviation fuel, heating oil, chemicals, and other products. About a quarter of the energy used by the industrial sector is for refining petroleum. Refineries today use about 30 percent less energy than they did in the 1970s, but rising fuel costs provide a challenge in maintaining reductions.

Steel Manufacturing

The steel industry uses energy to turn iron ore and scrap metal into steel. Hundreds of the products we use every day are made of steel. It is a very hard, durable metal and it must be heated to very high temperatures to manufacture it. Producing those high temperatures takes a lot of energy. The cost of energy in the steel industry is 15 percent of the total cost of making the steel. Most of this energy comes from coal and natural gas, or electricity generated from those sources.

Since 1990, the steel industry has reduced its energy consumption by 30 percent per ton of steel. New technology has made steel stronger so that less steel is needed for many uses. For example, the Willis Tower, formerly the Sears Tower, in Chicago could be built today using 35 percent less steel.

The use of recycled steel also saves energy. It requires 75 percent less energy to recycle steel than to make it from iron ore. Today, two-thirds of new steel is made from recycled scrap, making steel the nation's leading recycled product.

Aluminum Manufacturing

Aluminum is a very light-weight, versatile metal. We use aluminum to make soft drink cans, building materials, car parts, and many other products.

It takes huge amounts of electricity to make aluminum from bauxite, or aluminum ore. The cost for this energy is about one-third the total cost of manufacturing aluminum. Today, it takes 20 percent less energy to produce a pound of aluminum than it did 20 years ago. Using recycled aluminum requires about 95 percent less energy than converting bauxite into metal.

U.S. Industrial Energy Consumption by Source, 2015

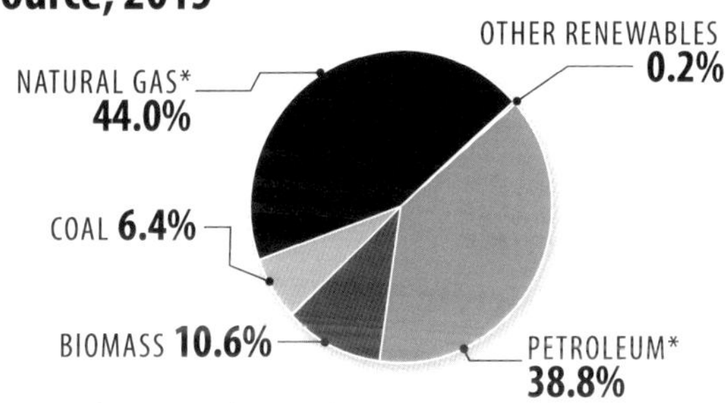

Data: Energy Information Administration
*Propane consumed by industry is included in petroleum and natural gas consumption figures.

Paper Manufacturing

The United States uses enormous amounts of paper every day— newspapers, books, bags, and boxes are all made of paper. Energy is used in every step of paper making. Energy is used to chop, grind, and cook the wood into pulp. More energy is used to roll and dry the pulp into paper.

The paper and pulp industry uses 30 percent less fossil fuels today than in the past, mainly because of better technology and increased use of wood waste to generate electricity on-site. Many industries have lowered energy use by using recycled materials. In the paper and pulp industry, it is not always cheaper to use recycled paper because it costs money to collect, sort, and process the waste paper.

Recycling has other benefits, though. It reduces the amount of paper in landfills and means fewer trees must be cut.

Chemical Manufacturing

Chemicals are an important part of our lives. We use chemicals in our medicines, cleaning products, fertilizers and plastics, as well as in many of our foods. Chemical manufacturing uses almost one-quarter of the energy consumed by the industrial sector.

The chemical industry uses energy in two ways. It uses coal, oil, and natural gas to power the machinery to make the chemicals. It also uses petroleum, propane, and natural gas as major sources of hydrocarbons from which the chemicals are made.

New technology has increased energy efficiency in the chemical industry by more than 50 percent in the last 35 years.

Cement Manufacturing

New roads and buildings are being built everywhere, every day. We use lots of concrete.

Concrete is made from cement, water, and crushed stone. A lot of energy is used in making cement. The process requires extremely high temperatures—up to 1,800 degrees Celsius (3,400°F).

Cement plants have reduced their energy consumption by more than one-third using innovative waste-to-energy programs. These wastes, such as printing inks, dry cleaning fluids, and used tires, have high energy content. For example, the energy content of one pound of tires is greater than one pound of coal. This industry is using energy that would otherwise be wasted in a landfill.

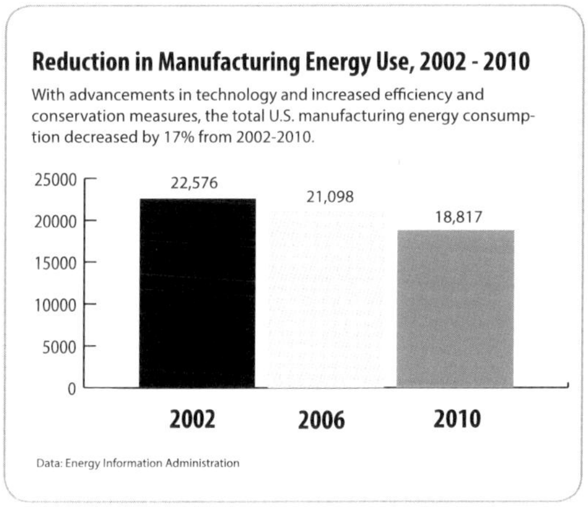

Transportation Sector Energy Consumption

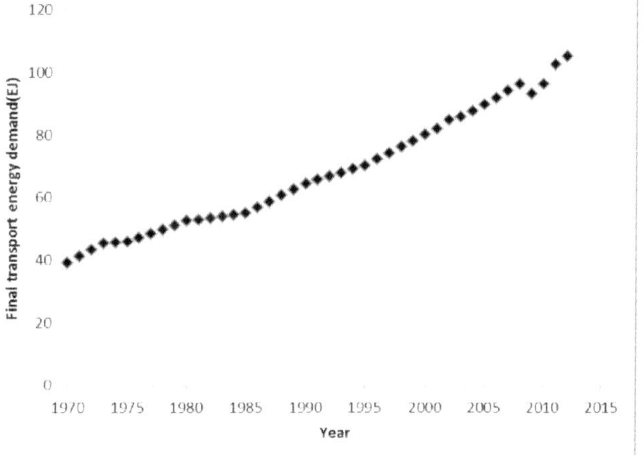

All motorized transport, whether for moving passengers or freight, requires the expenditure of energy. Even non-motorized transport requires the expenditure of food energy by humans, or fodder by animals. Today, transport energy uses nearly one-quarter of global primary energy. It is useful to distinguish between two categories of energy: primary and secondary. The International Energy Agency (IEA) defines total primary energy supply (TPES) as follows: 'production + imports – exports international marine bunkers international aviation bunkers ± stock changes. For the world total, international marine bunkers and international aviation bunkers are not subtracted from TPES'. Also, imports and exports are irrelevant at the global level.

Primary transport energy consumption thus includes not only the energy of the gasoline in the tank, for example, but also the energy needed to discover, produce, transport and refine the crude oil, and transport the gasoline to the service station. Following the total transport energy chain from oil wells to the motive power provided by the engine is often called a 'well-to-wheels energy' analysis.

However, vehicles do not use crude oil as fuel, but refined oil products such as gasoline, diesel and aviation turbine fuel. These refined fuels are called secondary energy and their combined use 'total final consumption' (TFC), to use the IEA terminology. The IEA defines TFC as: 'the sum of consumption by the different end-use sectors. Backflows from the petrochemical industry are not included in final consumption'. Thus in addition to petroleum-derived fuels, delivered natural gas, and electricity are also included in transport TFC. Energy analyses that just look at vehicle gasoline or diesel consumption are often termed 'tank-to-wheels' analyses, as in the familiar 'miles per gallon' (mpg) data for cars.

Vehicles engines consume their fuel in order to deliver transport services: passenger-km for pasenger transport and tonne-km for freight. Two people in a car who go on a 30 km trip generate 60 passenger-km, for example. For public transport, travel by drivers is excluded. For freight travel, only tonne-km by the payload is included, unless our interest is solely in engine efficiency. Road vehicles use up fuel to provide these services because they must overcome the 'road load'. This road load has three components: rolling resistance of the tyres, air resistance (drag), and inertial resistance the energy needed to increase the speed of a vehicle. Together with auxiliary power consumption of vehicles for lighting, air-conditioning, power steering etc, the road load accounts for a vehicle's total fuel use in liters or gallons of refined fuel.

Fuels used in Transport

Although coal was the original transport fuel for locomotive-powered trains, and electric motor and steam-powered road vehicles were popular at the turn of the century, oil-derived fuels have long dominated the energy mix for global transport. Table shows the breakdown in the various energy sources in final global energy demand in the transport sector. The table shows that use of both gas and biofuels has risen since 1973, with coal continuing its steady decline. For the dominant oil-derived fuels, diesel and aviation turbine fuels have both increased their share at the expense of gasoline. Nearly all trucks, buses and trains (except those using electric traction) run on diesel, and the share of diesel-fuelled passenger cars is also increasing,

Per liter of fuel, diesel vehicles are more efficient than petrol ones.

Table: Fuel shares of final world transport energy use in 1973 and 2012

Fuel	1973 (%)	2012 (%)
Oil	94.30	92.86
Coal	3.05	0.13
Gas	1.64	3.60
Electricity	0.98	1.02
Biofuels	0.03	2.39
Total	100.00	100.00

Reducing Transport Energy Consumption

Transport energy use can be cut in several possible ways:

- Increase the efficiency of the vehicle (vehicle-km per MJ of final energy)
- Increase the efficiency of conversion of primary into secondary energy
- Increase the vehicle loading rates
- Shift from less to more energy efficient transport modes.

The first option has received by far the most attention, but the last, modal shift, is now becoming an important option, especially for urban passenger transport.

Increase Efficiency of Vehicular Power Plant and Transmission

Two approaches are possible for increasing the vehicle-km/MJ (MJ = Megajoule = 106 J) for a given vehicle; increase the efficiency of the vehicle's power plant and transmission, or reduce the road load. The road load has three components—air friction, rolling friction and inertial load. The advantage of road load reductions is that the improvements can be applied to any vehicle, whereas engine efficiency improvements are often engine-specific; improvements to the combustion efficiency of a conventional car will be of little help in an all-electric vehicle.

According to Cullen, large energy savings are possible for cars by reductions in rolling friction (reducing both the coefficient of rolling friction and the vehicular body mass), in wind friction (reducing both the drag coefficient and the vehicle frontal area) and in inertial load (by reducing the vehicular body mass). Truck efficiency improvements can be made by similar methods.

These authors give the practical limit to energy efficiency improvements for each transport mode as follows:

- Cars 91%
- Trucks 54%
- Trains 57%
- Air 46%
- Ship 63%

- Weighted average 70%

The figures above are only for reducing the energy needs for transport vehicles. Still further energy savings are possible by engine and transmission improvements. In summary, we can both reduce the amount of fuel needed to produce a given power output, and we can also reduce the average power output needed to move the vehicle and its contents.

However, there are great difficulties in achieving these potential efficiency gains, as they can conflict with other desirable vehicle properties. For cars, high tire pressures reduce rolling friction but also ride comfort. Smaller, lighter, vehicles may have reduced safety, comfort, durability and performance. With smaller and lighter cars, the disparity in size and mass with heavy trucks sharing the same road space would further increase, with potential safety problems. Overall the gains in light vehicle on-road efficiency over recent decades has been disappointing. Increases in engine and transmission efficiency have been offset by the desire larger, higher-performance vehicles, and rising power needs for auxiliaries such as air-conditioning and power-steering.

Increase Efficiency of Conversion of Primary into Secondary Energy

Although transport efficiency is often measured on a vehicle-km/liter of fuel (or mpg) basis, the adverse effects from vehicular travel—oil depletion, CO_2 emissions, and air pollution—are the result of primary energy consumption. In other words, conversion energy losses cannot be ignored. Because of global depletion of conventional oil reserves, the world will need to rely increasingly on unconventional oil. Although the term is often restricted to heavy oil, shale oil and bituminous sands, ASPO extend the term to cover oil from deep water and from Polar Regions, and so unconventional oil is already an important share of total production. The justification is that these sources can also have high monetary and environmental costs.

Even for conventional oil, Murphy and Hall have shown that the ratio of energy output for a given level of input energy (the energy ratio) has fallen steadily over the years. For unconventional sources, the energy ratio will fall still further, and the CO_2 and environmental costs will rise alongside the monetary costs per liter of fuel delivered to the fuel tank. The result will be that any further energy efficiency improvements in vehicular tank-to-wheels efficiency will be at least partly offset by rising well-to-tank energy costs.

Increase Vehicle Loading Rates

Especially for OECD car travel, overall seat occupancy rates are low, typically about 1.5 persons per vehicle. For a 5-seater car, this gives a mere 30% seat occupancy rate. For public transport travel, occupancy rates vary enormously from country to country, and can be very high, not only in industrializing countries, but also in high-income Asian cities such as Tokyo and Hong Kong. Increasing occupancy rates is attractive because it can be achieved with the existing public transport or private vehicle stock. However, for private transport it is very difficult to implement, especially if household occupancy is falling, and car ownership and incomes are both rising. Raising car occupancy rates has proved difficult when tried because it entails car-sharing with non-family members, and much of the convenience of car travel over public transport, such as ability to choose one's own departure time, and travel privacy, would be lost.

It can also be difficult for public transport, but if public transport patronage is rising, as is the case in many OECD countries today, higher occupancy rates occur naturally. For many industrializing countries, public transport vehicles are often grossly overcrowded. Although this ensures both high energy and land use efficiency, loadings should be reduced in the interests of comfort and safety. For both public and private transport, occupancy rates vary by time of day and trip purpose. For public transport, loadings, and thus energy efficiency, are highest for the work trip, while for private travel, the opposite is the case, with peak hour congestion lowering fleet energy efficiency still further.

For air travel, loadings are already high, with little scope for further improvement. Freight transport typically has large payload to tare weight ratios compared with passenger transport. For specialized transport (bulk ore carriers, oil tankers etc.) back trips are usually made empty, but it is difficult to see how this could be changed.

Shift from Less to More Energy Efficient Transport Modes

The various transport modes can vary greatly in terms of primary energy efficiency, even with similar levels of loading. When differences in seat occupancy rates are allowed for, public transport modes are usually much more energy efficient than car travel, as can be seen most clearly by comparing bus and car travel, both of which use petroleum fuels. Comparisons of electric public transport (trains and trams) with car travel is complicated because of different efficiencies in primary to secondary energy conversion. Important reasons for the higher efficiency of rail travel compared with cars and trucks are the very low coefficient of rolling friction with steel wheels running on steel rails, and the lower air friction per unit load because of the small frontal area to length ratio.

The difference in primary energy efficiency is far more pronounced for freight modes, with sea freight up to two orders of magnitude more efficient (tonne-km/primary MJ) than air freight. The difficulty is that in terms of tonne-km, most world freight is already sent by energy-efficient water or rail modes. There is some scope for shifting some bulk freight from road to rail—if the necessary rail infrastructure is there—but far less for non-bulk freight or urban deliveries. Air freight accounts for only a tiny share of world tonne-km, and is used when delivery speed is at a premium.

How Much Do you use

In order to establish how your own electricity consumption compares with figures quoted previously you could

1. Check you own electricity bills over the period of a year to calculate how much electricity you use and how much it costs you.
2. Alternatively you could calculate your own electricity consumption figures using our Household Electricity Consumption Guide. This can help you calculate how much electricity you use in your house or individual rooms within the house - living room, kitchen, bedrooms / offices, and bathrooms.

The amount of energy a specific appliance consumes can be illustrated in the following formula:

Wattage x Hours Used Per Day

1000

= (1 kilowatt (kW) = 1,000 Watts) = Daily Kilowatt-hour (kWh) consumption

Multiply this by the number of days you use the appliance during the year for the annual consumption.

You can then calculate the annual cost to run an appliance by multiplying the number of kWh per year used by your electricity suppliers standard electricity charge per kWh consumed, typically around 6.8p/kWh.

Example:

1. If you watch TV 4 hours a day you can calculate how much electricity this will use per year.

TV Power rating = 100 Watts x 4 hours / day x 365 days / year

1000

= 146 kWh x 6.8 p / kWh

= £9.98 / year

2. Check how much electricity your personal computer and monitor use

Personal Computer and Monitor (220+150)

= 370 Watts x 4 hours/day x 365 days/year

1000

= 540 kWh x 6.8 p / kWh

= £36.73 / year

Appliances such as refrigerators, although turned "on" all the time, actually cycle on and off at a rate that depends on a number of factors. These factors include how well it is insulated, room temperature, freezer temperature, how often the door is opened, if the coils are clean, if it is defrosted regularly, and the condition of the door seals.

Electric Power Consumption

The total electric energy consumed by an individual (especially houses) measured in kilowatt-hour is called electric power consumption. All the power companies install a meter in their customer's house so that they can charge according to the electric consumption. Those installed meters measure electricity on kilowatt-hour (KWH). In normal terms, we say 1 unit for 1 KWH. It is actually the amount of power used for a certain period of time. For the easy purpose, we say an electrical appliance of 1000W consuming power for one hour will consume 1KWH i.e. 1KWH = 1KW x 1hr.

Power Consumption = Power consumed by appliance x no of appliance x time operating

= P (in KW) x N x T (in hour)

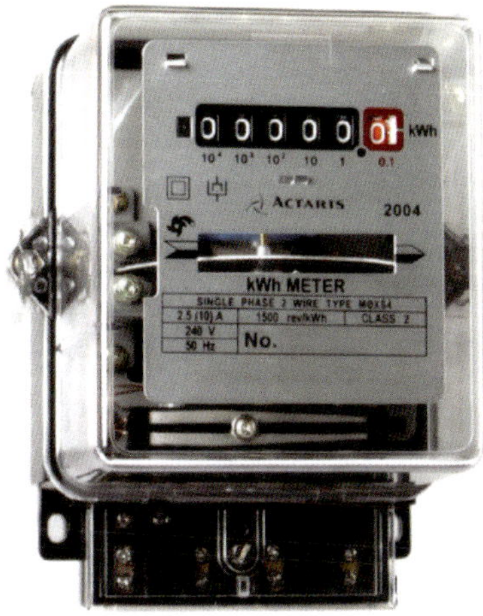

Amount to pay = power consumption x price per unit

So, to calculate total power consumption of a home, we need to know the power of each equipment, number of equipment and time they run. Since all the appliances are not run at the same time and same period of time it is better to calculate power consumed by each appliance and add all at last.

Example:

If a home has 10 tube lights of 30W that run for 4 hours daily, 5 television of 200W running for 3 hours daily, 2 irons 1000W 1 hour weekly. Calculate total amount to be paid if the cost per unit is Rs. 7.

Here,

Power consumed by tube lights in one day = 10 x 301000 x 4 = 65 KW Power consumed by tube lights in one month = 65 x 30 = 36 KW

Power consumed by television in one day = 5 x 2001000 x 3 = 3 KW Power consumed by television in one month = 3 x 30 = 90 KW

Power consumed by iron in one week = 2 x 10001000 x 1 = 2KW Power consumed by iron in one month = 2 x 4(one month has 4 weeks) = 8 KW

Total power consumption in one month = 36 + 90 + 8 = 134KW Total amount to be paid = total power consumption x price per unit= 134 x 7 = Rs. 938

Development Relevance: An economy's production and consumption of electricity are basic indicators of its size and level of development. Although a few countries export electric power, most production is for domestic consumption. Expanding the supply of electricity to meet the growing demand of increasingly urbanized and industrialized economies without incurring unacceptable social, economic, and environmental costs is one of the great challenges facing developing countries. Modern societies are becoming increasing dependent on reliable and secure electricity supplies to underpin economic growth and community prosperity. This reliance is set to grow as more efficient and less carbon intensive forms of power are developed and deployed to help decarbonize economies. Maintaining reliable and secure electricity services while seeking to rapidly decarbonize power systems is a key challenge for countries throughout the world. In developing economies growth in energy use is closely related to growth in the modern sectors - industry, motorized transport, and urban areas - but energy use also reflects climatic, geographic, and economic factors (such as the relative price of energy). Energy use has been growing rapidly in low- and middle-income economies, but high-income economies still use almost five times as much energy on a per capita basis. Governments in many countries are increasingly aware of the urgent need to make better use of the world's energy resources. Improved energy efficiency is often the most economic and readily available means of improving energy security and reducing greenhouse gas emissions.

Limitations and Exceptions: Data on electric power production and consumption are collected from national energy agencies by the International Energy Agency (IEA) and adjusted by the IEA to meet international definitions. Data are reported as net consumption as opposed to gross consumption. Net consumption excludes the energy consumed by the generating units. For all countries except the United States, total electric power consumption is equal total net electricity generation plus electricity imports minus electricity exports minus electricity distribution losses. The IEA makes these estimates in consultation with national statistical offices, oil companies, electric utilities, and national energy experts. The IEA occasionally revises its time series to reflect political changes, and energy statistics undergo continual changes in coverage or methodology as more detailed energy accounts become available. Breaks in series are therefore unavoidable.

Statistical Concept and Methodology: Electric power consumption per capita (kWh) is the production of power plants and combined heat and power plants less transmission, distribution, and transformation losses and own use by heat and power plants, divided by midyear population. Energy data are compiled by the International Energy Agency (IEA). IEA data for economies that are not members of the Organisation for Economic Co-operation and Development (OECD) are based on national energy data adjusted to conform to annual questionnaires completed by OECD member governments. Electricity consumption is equivalent to production less power plants' own use and transmission, distribution, and transformation losses less exports plus imports. It includes consumption by auxiliary stations, losses in transformers that are considered integral parts of those stations, and electricity produced by pumping installations. Where data are available, it covers electricity generated by primary sources of energy - coal, oil, gas, nuclear, hydro, geothermal, wind, tide and wave, and combustible renewables. Neither production nor consumption data capture the reliability of supplies, including breakdowns, load factors, and frequency of outages.

References

- Obrecht, Matevz; Denac, Matjaz (2013). "A sustainable energy policy for Slovenia : considering the potential of renewables and investment costs". Journal of renewable and sustainable energy. 5 (3): 032301. doi:10.1063/1.4811283.
- Energy-consumption-definition: dexma.com, Retrieved 12 March 2018
- "Project Closing Report. Natural Gas Centre of Excellence Project. Narrative" (PDF). 20 March 2005. Archived from the original (PDF) on 2007-09-27. Retrieved 2007-05-12
- Consumption-conservation, learning-center: saveonenergy.com, Retrieved 28 May 2018

- Robert Orttung and Indra Overland (2011) 'A Limited Toolbox: Explaining the Constraints on Russia's Foreign Energy Policy', Journal of Eurasian Studies, Vol. 2, No. 1, pp. 74-85
- World-energy-consumption-and-resources-An-outlook-for-the-rest-of-the-century-228622897: researchgate.net, Retrieved 21 June 2018
- Silvestre, B. S., Dalcol, P. R. T. Geographical proximity and innovation: Evidences from the Campos Basin oil & gas industrial agglomeration — Brazil. Technovation (2009), doi:10.1016/j.technovation.2009.01.003
- Global-Transport-Energy-Consumption-301650630: researchgate.net, Retrieved 15 July 2018
- The Energy of the Future: Fourth "Energy Transition" Monitoring Report — Summary(PDF). Berlin, Germany: Federal Ministry for Economic Affairs and Energy (BMWi). November 2015. Archived from the original (PDF) on 2016-09-20. Retrieved 2016-06-09

4

Energy Storage

The process of storing or capturing energy for future use is called energy storage. The topics elaborated in this chapter will help in gaining a better perspective about the different methods of energy storage such as pumped-storage, flywheel energy storage, battery energy storage system, power-to-gas, etc.

Energy storage is a mechanism to contain useful energy which can then be used at some later time in the future. It is a naturally occurring process, for example, fossil fuels are naturally created stores of the suns energy from millions of years ago. It is also a process that has been exploited by humans since they have existed – think of people gathering wood to burn to produce heat for cooking or heating.

Energy can be stored in many forms, but some are easier to contain than others. For example, it is difficult to store large amounts of energy as electricity, it is comparatively easier to an equivalent amount of chemical potential energy or even gravitational potential energy.

When the term 'Energy Storage' is used today it generally *excludes* fossil fuels, referring to a process by which potentially useful energy is absorbed at an earlier time and then released on demand at some useful time in the future *without* using traditional fuels like coal, gas, uranium etc. However it is important to realise that fuels (including traditional fuels like fossil fuels and uranium for nuclear fission as well as non-traditional fuels like hydrogen etc) are all forms of energy storage. The energy storage aspect is what makes them a fuel. At present the most important form of energy storage in most energy systems is fossil fuels – these are unmatched in terms of energy density, accessibility, low-cost, ease of handling and conversion and transportability.

Recent interest in energy storage has boomed due to the widespread uptake of renewable energy sources, which aim to reduce our dependency on fossil fuel generation. The motivation for this switch comes from the compelling evidence around climate change and its dangers to our species. As the electricity industry is the biggest source of global GHG emissions, renewable energy is an important method of mitigating climate change.

There is certainly enough renewable energy to power our global energy needs, especially if we engage in a significant amount of energy conservation and efficiency measures, however the biggest problem with renewable energy lies with its associated intermittency. That is to say that the wind does not blow all the time, nor does the sun shine all the time, nor even do the tides flow all the time and unfortunately we cannot store the wind as wind or the sunlight and sunlight (we can in a way store the tide as potential energy in water – tidal barrages do this – however this is probably a small contribution to our energy needs and id not without its drawbacks)! Further still, much of this intermittency is largely unpredictable; of the three examples listed only one renewable energy source is largely predictable – the tidal energy resource. Wind and solar can be predicted fairly accurately a few days before (with increasing accuracy as the forecast window to real time shortens) but beyond that we cannot tell when these energy resources will be available.

Energy storage is one way to tackle this intermittency issue head on. The idea is that by storing energy extracted from these renewable resources it can be available to use at a later more convenient point in time. Of course, for an entirely satisfactory solution it would be necessary to know that if we stored say a weeks' worth of energy the renewable resources would be available again sometime within that week to re-charge the energy storage before we ran out. It would also require that the energy storage had the power to charge and discharge at the necessary rates, and the capacity to store all the energy that we needed. That is the ideal – affordable large scale energy storage would be the "holy grail" for an entirely renewable based electricity system.

Approaches of Intermittency

There are a couple of different approaches to dealing with intermittency. One way that was heavily favoured over energy storage until recently was interconnection – the idea being essentially that we could rely on the geographical spread of renewable energy generators and compensate for reduced generation in one area by increased generation in another. Say that if it wasn't windy in the South of England it might be windy in the North of Scotland, or it may be sunny in the south of Wales (although South Wales residents may think this unlikely), so the energy from the area with favourable conditions would provide the whole system with renewable energy. This approach is very useful, but has two fundamental limitations. Firstly the geographical spread has to be very large and be very well connected via energy infrastructure and as it turns out, renewable resources are usually most energy dense where these is a sparse population and hence a weak grid infrastructure. Secondly this requires an impractical amount of over-capacity in many geographical locations.

Another approach is demand-side management. That is, instead of changing supply to match demand, demand can be reduced when supply is scarce. This is a great approach as it is often a very cheap solution (the cost of not using electricity at a particular time is often relatively low), however it requires an incentive for users to perform certain tasks at certain times, by making the price of electrical units (kWh) more expensive at certain times. For example, cheaper electricity at night encourages people to use electricity when the demand for electricity is usually lower. There are several existing schemes that already use this technique, and many electricity suppliers offer a two-tariff payment option meaning that electricity at night is cheaper than during the day. This is especially useful for technologies like storage-heating systems. The biggest problem with this approach is that in order to be effective consumers have to see real-time electricity prices, and be able to respond to them. There is significant resistance from legislators and the industry around this, and political concerns that some vulnerable consumers may suffer from very high bills due to price spikes. Crucially too, demand side management can only go so far, as most would consider it unacceptable if they could only turn on their television when the wind was blowing! Demand-side management can also be used to avoid asset enhancement.

Energy Storage Technologies

There are a number of ways energy storage can be achieved:

Compressed Air Storage

Used in conjunction with a wind farm, a compressed air energy storage system draws in air, creating a high-pressure system in a series of large underground chambers. When the wind speeds slows or power demand surges, the pressurized air, typically mixed with a bit of natural gas, is released to power turbines or generators.

Pumped-storage Hydroelectricity

A popular form of energy storage, pumped-storage hydroelectricity is typically designed to store excess power from the grid. When power demand lags, electricity from the grid is used to pump water up into a reservoir or lake. When demand surges, water is allowed to flow from the upper reservoir down to a lower reservoir. As it flows to a lower elevation, the water passes through turbines, generating electricity.

Advanced Rail Energy Storage

Devised by a California startup, Advanced Rail Energy Storage is a system that uses rail cars to reserve electricity. The unusual system uses excess grid energy to power axle-drive motors on rail cars, pushing mass uphill and against gravity to a storage yard.

To extract electricity from the system, the rail cars move back downhill. This time, the wheels acts as tiny generators, sending electricity back to the grid.

Flywheel Energy Storage

Flywheel energy storage systems store grid energy by converting electricity into kinetic energy in the form of spinning wheels. The wheels are held in a frictionless vacuum by a magnetic field to prevent a loss of energy, and when power is needed, the spinning can be slowed in a way that generates electricity.

This system offers a number of advantages over other energy storage systems, including low maintenance, long life, and low environmental impact.

Lithium-ion Battery Storage

Probably the easiest concept for people to understand, experts and officials are investing heavily in the development of large lithium-ion batteries to store power for when supply outpaces demand for electricity.

Lithium batteries weren't developed for industrial use and the widespread, grid-base adoption of lithium battery storage faces three major obstacles: cost; the fact that they can be a safety hazard; and their limited lifespan.

Liquid Air Energy Storage

Liquid air energy storage, or LAES, uses excess grid electricity to cool ambient air to the point it becomes a liquid. To extract electricity from the system, the liquid air is converted back to gas by exposure to ambient air or with waste heat. This expanding gas is then used to power turbines.

Pumped Heat Electrical Storage

Pumped heat electrical storage stockpiles electricity by using it to drive a heat pump that moves heat from a "cold store" to a hot store," comparable to the way a refrigerator moves heat from its interior. To recuperate the energy, the heat pump is flipped, taking from the hot store and putting it in the cold store. This generates mechanical work that drives a generator.

One of the advantages of this system over others is that it has a much faster response time than other grid storage systems, acting within minutes.

Redox Flow Batteries

Different from lithium-ion batteries, which are solid state batteries, redox flow batteries use chemical reduction and oxidation reactions to store excess grid energy in liquid electrolyte solutions.

The main advantage redox batteries have over lithium-ion and other solid state batteries is their flexibility. The power rating and energy rating on a redox flow battery can easily be modified for a given application by simply changing the stack size or size of storage tanks holding the electrolyte solutions.

Superconducting Magnetic Energy Storage

A superconducting magnetic energy storage system stores electricity not as chemical, kinetic, or potential energy, but as a magnetic field. The field is created by the flow of current in a superconducting coil that has been cooled to below a critical temperature point. The electromagnetic field requires very little energy to maintain, and energy is released by discharging the coils.

The advantage of this system is that it offers the capacity for instantaneous discharge.

Methane

German researchers have developed a way to take the products of biomass gasification to produce methane. As a gas that can then be burned to power turbines, methane essentially serves as an energy store medium in this scenario. Energy stored as methane could easily be piped or shipped using existing natural gas infrastructure.

Thermal Storage

Thermal storage is used for electricity *generation* by using power from the sun, even when the sun is not shining. Concentrating solar plants can capture heat from the sun and store the energy in water, molten salts, or other fluids. This stored energy is later used to generate electricity, enabling the use of solar energy even after sunset.

Concentrating solar plants focus the sun's heat to store energy in water, molten salts, or other fluids, which can be utilized even after the sun has set

Thermal storage technologies also exist for *end-use* energy storage. One method is freezing water at night using off-peak electricity, then releasing the stored cold energy from the ice to help with air conditioning during the day .

For example, Ice Energy's Ice Bear system creates a block of ice at night, and then uses the ice during the day to condense the air conditioning system's refrigerant . In this way, the Ice Bear system shifts the building's electricity consumption from the daytime peak to off-peak times when the electricity is less expensive.

Hydrogen

Hydrogen can be used as a zero-carbon fuel for *generation*. Excess electricity can be used to create hydrogen, which can be stored and used later in fuel cells, engines, or gas turbines to generate electricity without producing harmful emissions. NREL has studied the potential for creating hydrogen from wind power and storing it in the wind turbine towers for electricity generation when the wind isn't blowing .

The Wind to Hydrogen Project at NREL studies the storage of wind energy as hydrogen

Batteries

Batteries, like those in a flashlight or cell phone, can also be used to store energy on a large scale.

Like flywheels, batteries can be located anywhere so they are often seen as storage for *distribution*, when a battery facility is located near consumers to provide power stability; or *end-use*, like batteries in electric vehicles.

Batteries can be located in communities to provide power stability for homes

There are many different types of batteries that have large-scale energy storage potential, including sodium-sulfur, metal air, lithium ion, and lead-acid batteries. There are several battery installations at wind farms; which uses a 36 MW battery facility to help ensure stability of the power supply even when the wind isn't blowing .

Advancements in battery technologies have been made largely due to the expanding electric vehicle (EV) industry. As more developments are made with EVs, battery cost should continue to decline . Electric vehicles could also have an impact on energy storage through vehicle-to-grid technologies, in which their batteries can be connected to the grid and discharge power for others to use.

Future of Energy Storage

As new energy storage technologies are researched and tested, some barriers are likely to slow the commercialization of these technologies.

Research programs can help advance the deployment and commercialization of energy storage

Energy storage is expensive, especially without policies that place a monetary value on the unique benefits of storage. Plus there is no current need for additional storage capacity to maintain electricity grid reliability. Without an operational need, it is difficult for storage to be cost-effective in the present. Furthermore, storage lacks a robust track record of large commercial-scale projects (with the exception of pumped hydro), making it difficult to deploy new projects.

Despite these potential barriers, certain programs and policies can help drive the development and deployment of storage technologies.

The deployment of storage technologies can also be advanced through renewable electricity standards (RES). Some states recognize storage technologies as acceptable renewable generation in their RES, and other states award Renewable Energy Credits (REC) to energy generation from storage devices that were charged by renewables .

The Federal Energy Regulatory Commission (FERC), the agency that regulates the electricity grid, has created a pricing structure that pays storage technologies and other fast-ramping resources a higher price for their services. This pricing structure, called Pay-for-Performance, recognizes the value of rapid response in providing stability to the grid. Pay-for-Performance has the potential to make storage technologies more cost-effective on a commercial-scale. An investment tax credit (ITC) would also help accelerate the deployment of storage technologies.

With the support of government and industry, energy storage technologies can continue to develop and expand, aid in the increasing deployment of variable renewable energy sources, and help store an ever-growing amount of clean, renewable energy in the future.

Applications

Mills

The classic application before the industrial revolution was the control of waterways to drive water mills for processing grain or powering machinery. Complex systems of reservoirs and dams were constructed to store and release water (and the potential energy it contained) when required.

Home Energy Storage

Home energy storage is expected to become increasingly common given the growing importance of distributed generation of renewable energies (especially photovoltaics) and the important share of energy consumption in buildings. To exceed a self-sufficiency of 40% in a household equipped with photovoltaics, energy storage is needed. Multiple manufacturers produce rechargeable battery systems for storing energy, generally to hold surplus energy from home solar/wind generation. Today, for home energy storage, Li-ion batteries are preferable to lead-acid ones given their similar cost but much better performance.

Tesla Motors produces two models of the Tesla Powerwall. One is a 10 kWh weekly cycle version for backup applications and the other is a 7 kWh version for daily cycle applications. In 2016, a limited version of the Telsa Powerpack 2 cost $398(US)/kWh to store electricity worth 12.5 cents/kWh (US average grid price) making a positive return on investment doubtful unless electricity prices are higher than 30 cents/kWh.

Enphase Energy announced an integrated system that allows home users to store, monitor and manage electricity. The system stores 1.2 kWh hours of energy and 275W/500W power output.

Storing wind or solar energy using thermal energy storage though less flexible, is considerably less expensive than batteries. A simple 52-gallon electric water heater can store roughly 12 kWh of energy for supplementing hot water or space heating.

For purely financial purposes in areas where net metering is available, home generated electricity may be sold to the grid through a grid-tie inverter without the use of batteries for storage.

Grid Electricity and Power Stations

Renewable Energy Storage

Construction of the Salt Tanks which provide efficient thermal energy storage so that output can be provided after the sun goes down, and output can be scheduled to meet demand requirements. The 280 MW Solana Generating Station is designed to provide six hours of energy storage. This allows the plant to generate about 38 percent of its rated capacity over the course of a year

The 150 MW Andasol solar power station is a commercial parabolic trough solar thermal power plant, located in Spain. The Andasol plant uses tanks of molten salt to store solar energy so that it can continue generating electricity even when the sun isn't shining

The largest source and the greatest store of renewable energy is provided by hydroelectric dams. A large reservoir behind a dam can store enough water to average the annual flow of a river between dry and wet seasons. A very large reservoir can store enough water to average the flow of a river between dry and wet years. While a hydroelectric dam does not directly store energy from intermittent sources, it does balance the grid by lowering its output and retaining its water when power is generated by solar or wind. If wind or solar generation exceeds the regions hydroelectric capacity, then some additional source of energy will be needed.

Many renewable energy sources (notably solar and wind) produce variable power. Storage systems can level out the imbalances between supply and demand that this causes. Electricity must be used as it is generated or converted immediately into storable forms.

The main method of electrical grid storage is pumped-storage hydroelectricity. Areas of the world such as Norway, Wales, Japan and the US have used elevated geographic features for reservoirs, using electrically powered pumps to fill them. When needed, the water passes through generators and converts the gravitational potential of the falling water into electricity. Pumped storage in Norway, which gets almost all its electricity from hydro, has currently a capacity of 1.4 GW but since the total installed capacity is nearly 32 GW and 75% of that is regulable, it can be expanded significantly.

Some forms of storage that produce electricity include pumped-storage hydroelectric dams, rechargeable batteries, thermal storage including molten salts which can efficiently store and release very large quantities of heat energy, and compressed air energy storage, flywheels, cryogenic systems and superconducting magnetic coils.

Surplus power can also be converted into methane (sabatier process) with stockage in the natural gas network.

In 2011, the Bonneville Power Administration in Northwestern United States created an experimental program to absorb excess wind and hydro power generated at night or during stormy periods that are accompanied by high winds. Under central control, home appliances absorb surplus energy by heating ceramic bricks in special space heaters to hundreds of degrees and by boosting the temperature of modified hot water heater tanks. After charging, the appliances provide home heating and hot water as needed. The experimental system was created as a result of a severe 2010 storm that overproduced renewable energy to the extent that all conventional power sources were shut down, or in the case of a nuclear power plant, reduced to its lowest possible operating level, leaving a large area running almost completely on renewable energy.

Another advanced method used at the former Solar Two project in the United States and the Solar Tres Power Tower in Spain uses molten salt to store thermal energy captured from the sun and then convert it and dispatch it as electrical power. The system pumps molten salt through a tower or other special conduits to be heated by the sun. Insulated tanks store the solution. Electricity is produced by turning water to steam that is fed to turbines.

Since the early 21st century batteries have been applied to utility scale load-leveling and frequency regulation capabilities.

In vehicle-to-grid storage, electric vehicles that are plugged into the energy grid can deliver stored electrical energy from their batteries into the grid when needed.

Air Conditioning

Thermal energy storage (TES) can be used for air conditioning. It is most widely used for cooling single large buildings and/or groups of smaller buildings. Commercial air conditioning systems are the biggest contributors to peak electrical loads. In 2009, thermal storage was used in over 3,300 buildings in over 35 countries. It works by creating ice at night and using the ice to for cooling during the hotter daytime periods.

The most popular technique is ice storage, which requires less space than water and is less costly than fuel cells or flywheels. In this application, a standard chiller runs at night to produce an ice pile. Water then circulates through the pile during the day to chill water that would normally be the chiller's daytime output.

A partial storage system minimizes capital investment by running the chillers nearly 24 hours a day. At night, they produce ice for storage and during the day they chill water. Water circulating through the melting ice augments the production of chilled water. Such a system makes ice for 16 to 18 hours a day and melts ice for six hours a day. Capital expenditures are reduced because the chillers can be just 40 - 50% of the size needed for a conventional, no-storage design. Storage sufficient to store half a day's available heat is usually adequate.

A full storage system shuts off the chillers during peak load hours. Capital costs are higher, as such a system requires larger chillers and a larger ice storage system.

This ice is produced when electrical utility rates are lower. Off-peak cooling systems can lower energy costs. The U.S. Green Building Council has developed the Leadership in Energy and Environmental Design (LEED) program to encourage the design of reduced-environmental impact buildings. Off-peak cooling may help toward LEED Certification.

Thermal storage for heating is less common than for cooling. An example of thermal storage is storing solar heat to be used for heating at night.

Latent heat can also be stored in technical phase change materials (PCMs). These can be encapsulated in wall and ceiling panels, to moderate room temperatures.

Transport

Liquid hydrocarbon fuels are the most commonly used forms of energy storage for use in transportation, followed by a growing use of Battery Electric Vehicles and Hybrid Electric Vehicles. Other energy carriers such as hydrogen can be used to avoid producing greenhouse gases.

Public transport systems like trams and trolleybuses require electricity, but due to their variability in movement, a steady supply of electricity via renewable energy is challenging. Photovoltaic systems installed on the roofs of buildings can be used to power public transportation systems during periods in which there is increased demand for electricity and access to other forms of energy are not readily available.

Electronics

Capacitors are widely used in electronic circuits for blocking direct current while allowing alternating current to pass. In analog filter networks, they smooth the output of power supplies. In resonant circuits they tune radios to particular frequencies. In electric power transmission systems they stabilize voltage and power flow.

Use Cases

The United States Department of Energy International Energy Storage Database (IESDB), is a free-access database of energy storage projects and policies funded by the United States Department of Energy Office of Electricity and Sandia National Labs.

Storage Capacity

Storage capacity is the amount of energy extracted from a power plant energy storage system; usually measured in joules or kilowatt-hours and their multiples, it may be given in number of hours of electricity production at power plant nameplate capacity; when storage is of primary type (i.e., thermal or pumped-water), output is sourced only with the power plant embedded storage system.

Benefits of Energy Storage

Because some renewable energy technologies – such as wind and solar – have variable outputs, storage technologies have great potential for smoothing out the electricity supply from these sources and ensuring that the supply of generation matches the demand.

Energy storage can play an important role in creating a more flexible and reliable electricity system

Energy storage is also valued for its rapid response – most storage technologies can begin discharging power to the grid very quickly, while fossil fuel sources tend to take longer to ramp up. This rapid response is important for ensuring stability of the grid when unexpected increases in demand occur.

Energy storage also becomes more important the farther you are from the electrical grid. For example, when you turn on the lights in your home, the power comes from the grid; but when you turn on a flashlight while camping, you must rely on the stored energy in the batteries. Similarly, homes that are farther away from the transmission grid are more vulnerable to disruption than homes in large metropolitan areas. Islands and microgrids that are disconnected from the larger electrical grid system depend on energy storage to ensure power stability, just like you depend on the batteries in your flashlight while camping.

Pumped-storage

Pumped storage is the process of storing energy by using two vertically separated water reservoirs.

Water is pumped from the lower reservoir up into a holding reservoir. Pumped storage facilities store excess energy as gravitational potential energy of water. Since these reservoirs hold such large volumes of water, pumped water storage is considered to be a large scale energy storage system. These pumped storage facilities are moderately efficient, with a round-trip efficiency of about 65-70%. The amount of energy stored depends on the mass of water pumped and the height difference between the reservoirs.

Pumped storage is a dispatchable source of energy since it can be deployed whenever demand is needed. It is often used to meet demand when intermittent, non-dispatchable sources, such as wind and solar power, cannot do so.

Currently pumped storage has the highest capacity of energy storage on the grid and accounts for 99% of bulk storage capacity in the world. The United States alone has more than 20 GW of pumped storage capacity today with facilities in every region of the country.

Over the past 70 years, there has been a steady increase in the number of pumped storage facilities around the world.

Operation

In periods of low demand, water is pumped up into the reservoir, generally using some sort of reversible turbine/pump such as a Francis turbine. In these low demand times, excess electricity from the grid is used to pump the water up. When electricity is in high demand, hydropower is generated by releasing the water stored in the reservoir. This water flows down a slope through penstocks, then flows through turbines, generating electricity.

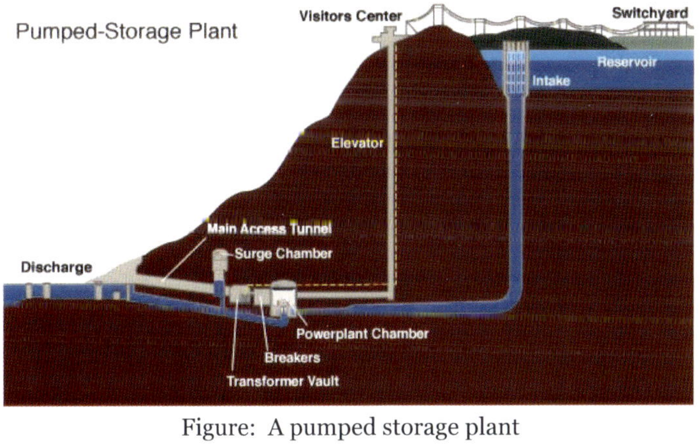

Figure: A pumped storage plant

Currently, the implementation of saltwater pumped storage and underground pumped storage facilities are being researched. These ideas are still a work-in-progress and implementing these facilities has been found difficult because of corrosion, as well as groundwater and surface water contamination issues and overcoming possibly higher operation and management costs. Okinawa Pumped Storage Plant is currently the only operational saltwater pumped storage plant in the world. The plant has an energy capacity of 30 MW with an elevation of around 154 meters and an effective head of 136 meters (the difference of 18 m is a way to describe the inefficiencies in the system). The plant uses paint-coated carbon steel with cathodic protection in order to prevent corrosion.

It is important to note that pumped storage facilities are net consumers of energy. This is a result of the energy lost pumping the water up into the reservoir. However, pumped storage is economical because of a net increase in revenue. This is because the electricity used to pump the water is less expensive than the electricity sold at the time of peak energy demand.

Feasibility of Technology and Operational Necessities

Conventional Pumped Hydro Storage

Conventional pumped hydro uses two water reservoirs, separated vertically. During off peak hours water is pumped from the lower reservoir to the upper reservoir. When required, the water flow is reversed to generate electricity. Some high dam hydro plants have a storage capability and can be dispatched as a pumped hydro. Underground pumped storage, using flooded mine shafts or other cavities, are also technically possible. Open sea can also be used as the lower reservoir. A seawater pumped hydro plant was first built in Japan in 1999.

While pumped hydro is often performed in geologically appropriate areas, in other words areas which have high elevation differences, pumped hydro can be constructed in such a way that it is applicable anywhere where water is available. An example of such a construction is shown in figure. However, the construction of such facilities is more expensive as absence of the natural vertical difference has to overcome due to the artificial construction of vertical difference.

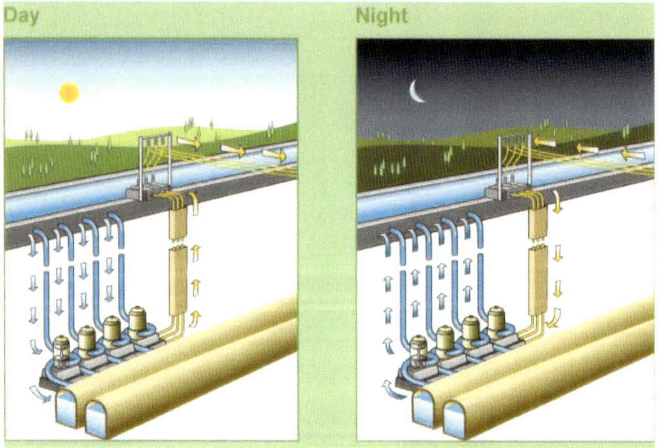

Figure: Another construction method for pumped hydro. This method is less dependent on naturally occuring geological conditions

Pumped Hydro with Variable Speed Turbines for Bulk Storage

A variation on the conventional pumped hydroelectric plant is based on the use of adjustable- or variable-speed turbines. An adjustable-speed pumped hydro plant is based on a turbine and motor generator that will operate over a range of rotation speeds that is 10% above and below nominal. This speed variation is accomplished by using a solid rotor with a three-phase winding that can be driven at a variable frequency. A cycloconverter changes AC power to the appropriate frequency via an AC-DC-AC converter. The stator, as in a conventional pumped hydro plant, is directly connected to the three-phase bus and is energized at line frequency. The modest ±10% change in rotation frequency has a significant impact on performance. The 20% variation in rotational velocity translates into a 50% change in power output because mass flow is related to the third power of the rotational speed.

An adjustable-speed pumped hydro plant has several advantages:

1. There is no need for a pony motor to start pumping.
2. Synchronous power operation covers a wider range than for a conventional turbine, from below 70% to full power.
3. The rate of changing the power output is driven by inertia of the water flow and is much more rapid than conventional pumped hydro.
4. Full power output can be delivered from water-head variations of a factor of two.
5. Rotational speed can be adjusted to avoid resonances within the equipment and cavitation modes in the water flow. This leads to longer life and less maintenance.
6. Higher overall efficiency, as high as a 0.3 improvement on an annual basis is possible.
7. Speed regulation provides frequency regulation to the grid.

However, improved performance comes with extra design and construction issues and costs:

1. Motor/generator balance is more critical and takes multiple adjustments prior to initial operation.
2. Higher speed rotation >350 rpm is required to obtain maximum performance.
3. The motor/generator/turbine shaft is longer, which requires more excavation, increasing civil engineering costs.
4. There is some additional cost for the cyclo-converter, but it is partially offset by the electronic conversion needed for starting conventional plants

Several of these plants are now installed in Japan and others are in design or under construction. The general philosophy of recent pumped-hydro plants has been to install equal numbers of adjustable speed turbines and conventional turbines. This approach seems to provide adequate flexibility for operation over a wide range of pumping and generating scenarios.

Typically, upper and lower reservoirs have been constructed by dams. Within the last 10 years, advanced pumped storage (APS) technology has been developed to improve speed, reliability and efficiency. These plants are designed hydraulically and mechanically for ultrafast loading and ramping, allowing frequent and rapid (<15 sec) changes among the pumping, generating and stand-by spinning modes.

Pump-back Hydroelectric Dams

Conventional hydroelectric dams may also make use of pumped storage in a hybrid system that both generates power from water naturally flowing into the reservoir as well as storing water pumped back to the reservoir from below the dam. The Grand Coulee Dam in the US was expanded with a pump-back system in 1973. Existing dams may be repowered with reversing turbines thereby extending the length of time the plant can operate at capacity. Optionally a pump back powerhouse such as the Russell Dam may be added to a dam for increased generating capacity. Making use of an existing dams upper reservoir and transmission system can expedite projects and reduce costs.

Potential Technologies

Seawater

Pumped storage plants can operate with seawater, although there are additional challenges compared to using fresh water. In 1999, the 30 MW Yanbaru project in Okinawa was the first demonstration of seawater pumped storage. It has since been decommissioned. A 300 MW seawater-based Lanai Pumped Storage Project was considered for Lanai, Hawaii, and seawater-based projects have been proposed in Ireland. A pair of proposed projects in the Atacama Desert in northern Chile would use 600 MW of photovoltaic solar (Skies of Tarapacá) together with 300 MW of pumped storage (Mirror of Tarapacá) raising seawater 600 metres (2,000 ft) up a coastal cliff.

Underground Reservoirs

The use of underground reservoirs has been investigated. Recent examples include the proposed Summit project in Norton, Ohio, the proposed Maysville project in Kentucky (underground limestone mine), and the Mount Hope project in New Jersey, which was to have used a former iron mine as the lower reservoir. The proposed energy storage at the Callio site in Pyhäjärvi (Finland) would utilize the deepest base metal mine in Europe, with 1,450 metres (4,760 ft) elevation difference. Several new underground pumped storage projects have been proposed. Cost-per-kilowatt estimates for these projects can be lower than for surface projects if they use existing underground mine space. There are limited opportunities involving suitable underground space, but the number of underground pumped storage opportunities may increase if abandoned coal mines prove suitable.

In Bendigo, Victoria, Australia, the Bendigo Sustainability Group has proposed the use of the old gold mines under Bendigo for Pumped Hydro Energy Storage. Bendigo has the greatest concentration of deep shaft hard rock mines anywhere in the world with over 5,000 shafts sunk under Bendigo in the second half of the 19th Century. The deepest shaft extends 1,406 metres vertically underground. A recent pre-feasibility study has shown the concept to be viable with a generation capacity of 30 MW and a run time of 6 hours using a water head of over 750 metres.

Decentralised Systems

Small pumped-storage hydropower plants can be built on streams and within infrastructures, such as drinking water networks and artificial snow making infrastructures. Such plants provide distributed energy storage and distributed flexible electricity production and can contribute to the decentralized integration of intermittent renewable energy technologies, such as wind power and solar power. Reservoirs that can be used for small pumped-storage hydropower plants could include natural or artificial lakes, reservoirs within other structures such as irrigation, or unused portions of mines or underground military installations. In Switzerland one study suggested that the total installed capacity of small pumped-storage hydropower plants in 2011 could be increased by 3 to 9 times by providing adequate policy instruments.

Underwater Reservoirs

In March 2017, the research project StEnSea (Storing Energy at Sea) announced their successful completion of a four-week test of a pumped storage underwater reservoir. In this configuration a hollow sphere submerged and anchored at great depth acts as the lower reservoir, while the upper reservoir is the enclosing body of water. Electricity is created when water is let in via a reversible turbine integrated into the sphere. During off-peak hours the turbine changes direction and pumps the water out again, using "surplus" electricity from the grid. The quantity of power created when water is let in grows proportionally to the height of the column of water above the sphere, in other words: the deeper the sphere is located, the more potential energy it can store, which can be transformed into electric power. On the other hand, pumping the water back out at greater depths also uses up more power, since the turbine-turned-pump must act on the same entire column of water.

As such the energy storage capacity of the submerged reservoir is not governed by the gravitational energy in the traditional sense, but rather by the vertical pressure variation.

While StEnSea's test took place at a depth of 100 m in the fresh water Lake Constance, the technology is foreseen to be used in salt water at greater depths. Since the submerged reservoir needs only a connecting electrical cable, the depth at which it can be employed is limited only by the depth at which the turbine can function, currently limited to 700 m. The challenge of designing salt water pumped storage in this underwater configuration brings a range of advantages:

- No land area is required,
- No mechanical structure other than the electrical cable needs to span the distance of the potential energy difference,
- In the presence of sufficient seabed area multiple reservoirs can scale the storage capacity without limits,
- Should a reservoir collapse, the consequences would be limited apart from the loss of the reservoir itself,
- Evaporation from the upper reservoir has no effect on the energy conversion efficiency,
- Transmission of electricity between the reservoir and the grid can be established from a nearby offshore wind farm limiting transmission loss and obviating the need for onshore cabling permits.

A current commercial design featuring a sphere with an inner diameter of 30 m submerged to 700 m would correspond to a 20 MWh capacity which with a 5 MW turbine would lead to a 4-hour discharge time. An energy park with multiple such reservoirs would bring the storage cost to around a few eurocents per kWh with construction and equipment costs in the range €1,200-€1,400 per kW. To avoid excessive transmission cost and loss, the reservoirs should be placed off deep water coasts of densely populated areas, such as Norway, Spain, USA and Japan. With this limitation the concept would allow for worldwide electricity storage of close to 900 GWh.

For comparison, a traditional, gravity-based pumped storage capable of storing 20 MWh in a water reservoir the size of a 30 m sphere would need a hydraulic head of 519 m with the elevation spanned by a pressurized water pipe requiring typically a hill or mountain for support.

Home use

Figure: Pico Hydro

Using a pumped-storage system of cisterns and small generators, pico hydro may also be effective for “closed loop” home energy generation systems.

PHS Performance Characteristics and Applications

Pumped hydro storage is at present the only utilised method of large-scale grid energy storage. It comprises 3% of global installed power capacity installed and 97% of global installed electrical energy storage. PHS has a time response from a few seconds to a few minutes, a large scale output and storage capacity with hourly to daily output durations. It can be used to provide substantial benefits to the energy system including frequency control, ramping/load levelling and peak shaving, load following, and provision of stand by reserve. Due to the low energy density of pumped storage schemes they are really only applicable for large scale grid applications.

Superconducting Magnetic Energy Storage

SMES was first proposed as an energy storage technology for power systems. SMES systems have fast response and high efficiency. Hence, electric utilities and military are focusing their attention towards SMES system. When compared with other energy storage technologies, SMES systems are costly. However, the integration of an SMES coil into the FACTS devices eliminates the cost for the inverter unit, which is the largest portion of the cost for the entire SMES system. Recently, high temperature superconductors are in the research stage. Such type is cost effective due to the reductions in refrigeration needs. There are a number of ongoing SMES projects currently installed or in development throughout the world.

In the SMES system, the energy is stored in the magnetic field generated by the dc current flowing through superconducting coil. The inductively stored energy (E in joules) and rated power (P in watts) are the commonly given specifications for SMES devices. The energy and power are expressed as in equation below.

$$E = \frac{1}{2}LI^2; P = \frac{dE}{dt} = LI\frac{dI}{dt} = VI$$

where 'L' is the inductance of the superconducting coil, 'I' is the dc current flowing through the coil and V is the voltage across the coil. The energy is stored as circulating current. The energy can be drawn from an SMES unit with almost instantaneous response with energy stored or delivered over periods ranging from a fraction of a second to several hours.

An SMES unit consists of a large superconducting coil at the cryogenic temperature. Figure below represents the block diagram of the SMES system. The superconducting coil is maintained at cryogenic temperature. This temperature is maintained by a cryostat or dewar that contains helium or nitrogen liquid vessels. A power conversion/conditioning system connects the SMES unit to an ac power system and it is used to charge/discharge the coil. There are two types of power conversion systems. First one uses a Current Source Converter (CSC) to both interface to the ac system and charge/discharge the coil and the second type uses a Voltage Source Converter (VSC) to interface to the ac system and a dc-dc chopper to charge/discharge the coil. The VSC and dc-dc chopper share a common dc bus. The SMES coil is charged or discharged by applying a positive or negative voltage across the superconducting coil. The SMES system enters a standby mode operation when the average voltage across the coil is zero, resulting in a constant average coil current.

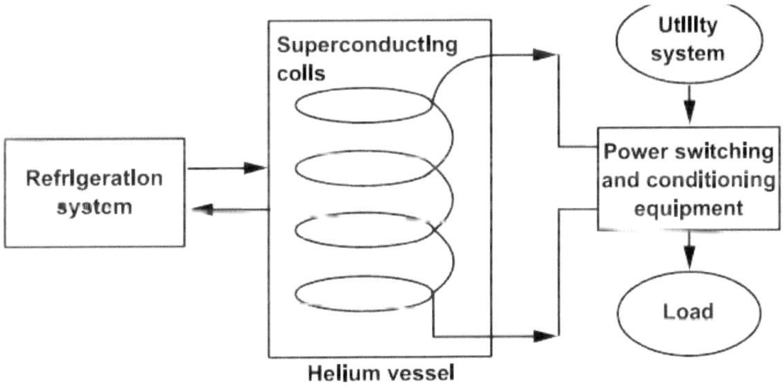

Figure: Block diagram of SMES system

Several factors such as coil configuration, energy capability, structure and operating temperature are taken into account in the design of the coil to achieve the best possible performance of an SMES system at the least cost. A compromise is made between each factor considering the parameters of energy/mass ratio, stray magnetic field and minimizing the losses for a reliable, stable and economic SMES system. The coil configuration can be a solenoid or a toroid. The solenoid type is simple and cost effective. The operating temperature used for a superconducting device is a compromise between the cost and the operational requirements.

It has been considered for load levelling, frequency support during loss of generation, enhancing transient and dynamic stability, dynamic voltage support, improving power quality and increasing transmission line capacity, thus enhancing the overall reliability of power systems. The energy power characteristics for potential SMES applications for generation, transmission and distribution are shown in figure below. The square area in the figure represents the applications that are currently economical. Therefore, the SMES technology has a unique advantage in two types of application: power system transmission control and stabilization and power quality.

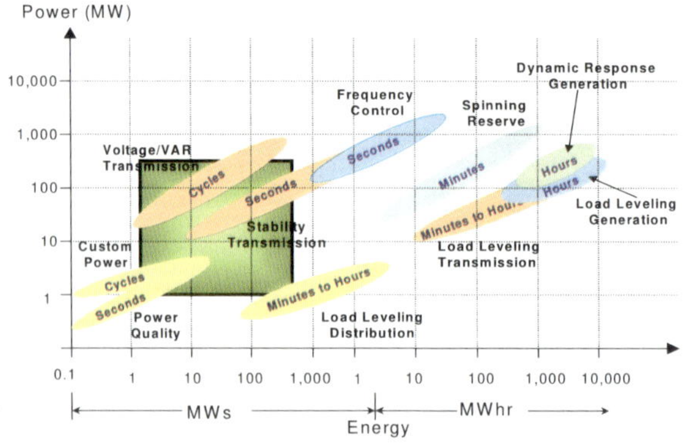

Figure: Energy-power characteristics for potential SMES applications for generation, transmission and distribution

SMES System Coil Modelling

The methods available for the determination of the coil characteristics are direct measurement on actual winding, direct measurements on an electromagnetic model of the coil or building a mathematical model and determining the voltage distribution and frequency response by means of the computer aided analysis. The first two methods, give more reliable and accurate results than the mathematical analysis. But, from the economical point of view, setting up of a mathematical model is the most convenient method with the lowest cost. Also, many design alternatives and detailed analysis can be done with the mathematical model of a winding.

A lumped parameter network model normally used and it contains magnetic and dielectric circuits. The magnetic circuit is represented by self and mutual inductance (L and M) of each turn. The dielectric circuit is represented by capacitances between adjacent turns (C_{adj}), axially separated turns (C_{ax}) and turn to the outside surface (C_g). Due to the high memory and computing costs, various degrees of simplification are necessary. When a superconducting coil is simulated for a purpose of dynamic operation, it is a common practice to represent the coil as an inductor.

Integration of SSSC with SMES System

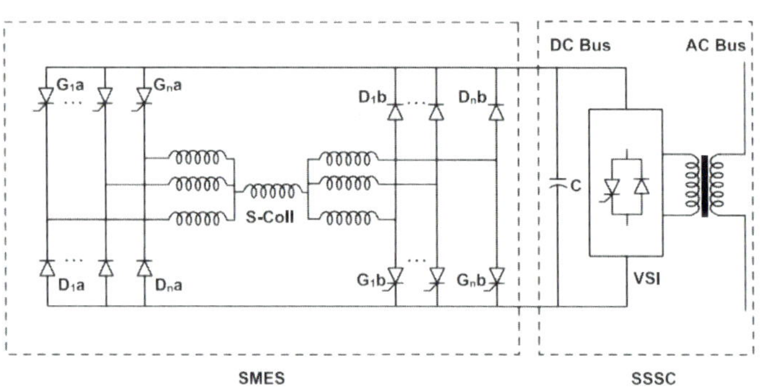

Figure: Structure of a GTO based chopper

For the integration of SSSC with SMES, a multiphase GTO chopper is modelled as shown in figure below. In FACTS / SMES application, the dc-dc chopper is connected to a FACTS device through a link capacitor maintaining a constant dc voltage.

Operating principles of the multiphase dc chopper can be explained with the help of a single phase dc chopper. In a single phase chopper, the GTO firing signal is a square wave with a specified duty cycle. The average voltage and current of the SMES coil are related to the dc source voltage and current of the SMES coil are related to the dc source voltage and average current by the duty cycle applied. These relationships can be expressed as,

$$V_{SME_{av}} = [1 - 2d] V_{dcav}$$

$$I_{dcav} = [1 - 2d] I_{SMES_{av}}$$

where, $V_{SME_{av}}$ is the average voltage across the SMES coil and $I_{SMES_{av}}$ is the average current through the SMES coil. V_{dcav} is the average dc source voltage, I_{dcav} is the average dc source current and d is the duty cycle of the chopper.

There is no energy transfer at a duty cycle of 0.5, where the average voltage of the coil is zero and the average coil current is constant. In cases of duty, when cycle being larger than 0.5 or less than 0.5, the coil is either charging or discharging respectively. Adjusting the duty cycle of the GTO firing signals controls the rate of charging/discharging. For an n-phase dc chopper, the duty cycle of firing signals of each phase is 1/n of the total duty cycle. In the three phase chopper used, the duty cycle of each phase changes from 0 to 1/3, and the frequencies of the GTO firing signals are 100Hz.

The dc-dc chopper has three modes of operation, in order to perform the charge, discharge and the storage in the SMES device. The chopper is operated in the step-down (Buck) configuration in the charge mode of the superconducting coil. Here, the set of thyristors "a" are operated with the duty cycle D while the set of thyristors "b" are kept "ON" all times. The relationship between the coil voltage and the dc bus voltage is stated by equation below.

$$V_{SMES} = D.V_{dc}$$

Once the charging of the superconducting coil is completed, the operating mode of the dc-dc converter is changed to the stand-by mode for which the set of thyristors "a" are kept "OFF" all the time while the set of thyristors "b" are kept "ON" constantly. Thus, the set of diodes "a", also called free-wheeling diodes, permits the stored current to flow continuously until it is required (neglecting losses at semiconductors).

Finally, in the discharge mode the chopper is operated in a step-up (Boost) configuration. Here, the set of thyristors "b" is operated with duty cycle D and while the set of thyristors "a" is kept "OFF" at all times. It should be remarked that the capacitor of the SSSC also works like part of the dc-dc chopper to increase the voltage applied by the SMES device to the dc bus. The relationship between the coil voltage and the dc bus voltage is stated by equation below.

$$-V_{SMES} = (1 - D).V_{dc}$$

The dc-dc chopper can be operated in a continuous spectrum of duty cycle values ranging from 0 to 1. The voltage in the dc bus is related to the output voltage of the VSI of the SSSC system through the equation below:

$$V_{dc} = k_a \cdot |V_{inv}|$$

Where k_a = ka

k is a constant associated with the pulse number of the VSI.

For the 48 pulse VSI, $k = 2\frac{\sqrt{2}}{\pi}$ and

a is the voltage ratio of the coupling transformer.

The total voltage ratio of the zigzag phase-shifting transformers is = $4\frac{n_2}{n_1}$

Internal Control Scheme for the Integrated SSSC - SMES System

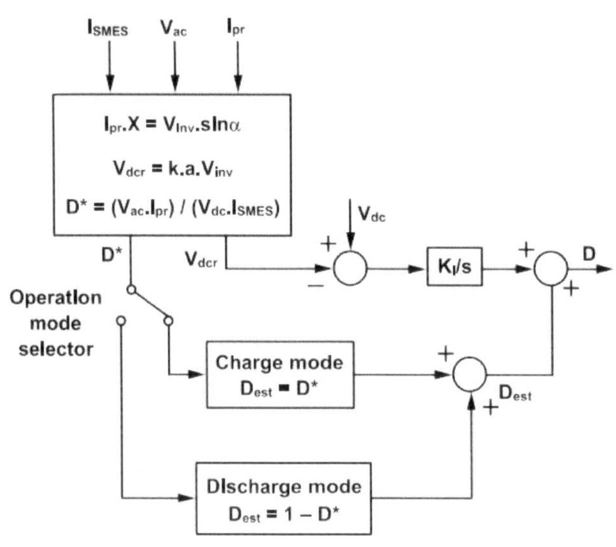

Figure: Control block diagram for the duty cycle of dc/dc chopper

A simplified functional scheme of the internal control of the SSSC/SMES system is shown in figure above. The control scheme that is proposed coordinates the control subsystems of the chopper and the VSI of the SSSC. The control scheme of the dc-dc chopper uses the control algorithms proposed by Lassetter and Jalali and Marcelo G.Molina et al as the basis. The control of the inverter is of the decoupled type according to the components of active and reactive power. This control has independent inputs of the reference signals of the current injected in the connection bus, that is to say, the component of the wanted reactive current I_{qr} and the component of the wanted active current I_{pr}.

The duty cycle D is estimated from the active power ratings that the SSSC should inject ($V_{ac}.I_{pr}$), from the voltage at the dc bus (V_{dc}) and from the current stored into the superconducting magnetic coil I_{SMES}. This estimated value (D^*_{est}) is adjusted through a closed loop control whose function is to eliminate the voltage error between the calculated and the real voltage ratings at the dc bus through an integral controller. The operating mode selector selects the proper mode for the charging and discharging of the SMES coil. The value of K_i is set through a trial and error method and the value is fixed as 25.

Simulation Results

For the test system, SSSC/SMES combination is inserted. Simulation is carried out for the power flow control initially. The real power reference is initially set to 8.5 pu and then at 0.25 sec it is increased to 13 pu. Since the SMES unit can exchange the real power demand of the system, the 13 pu is achieved. Figure below shows the waveform for the real power control of the SSSC/SMES integrated system.

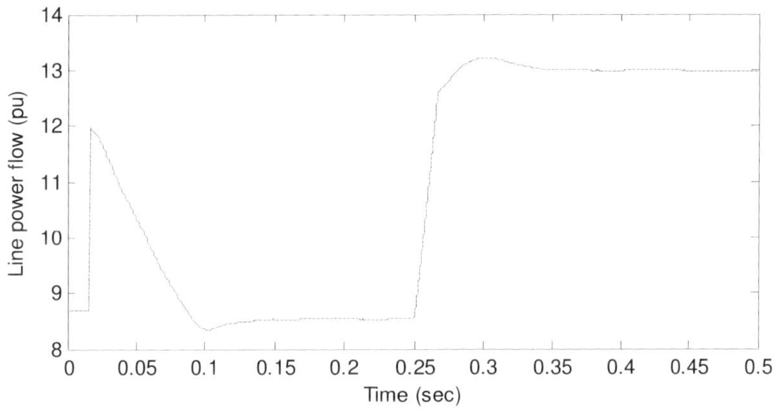

Figure: Real power flow through transmission line 2

Figure below shows the reactive power control of the SSSC-SMES integrated system. The real power reference is fixed to 11 pu and the reactive power reference is changed from 0.5 pu to 0.8 pu at 0.25 sec. Compared to the system with SSSC in the integrated SSSC/SMES system, the peak overshoot and the oscillations are reduced.

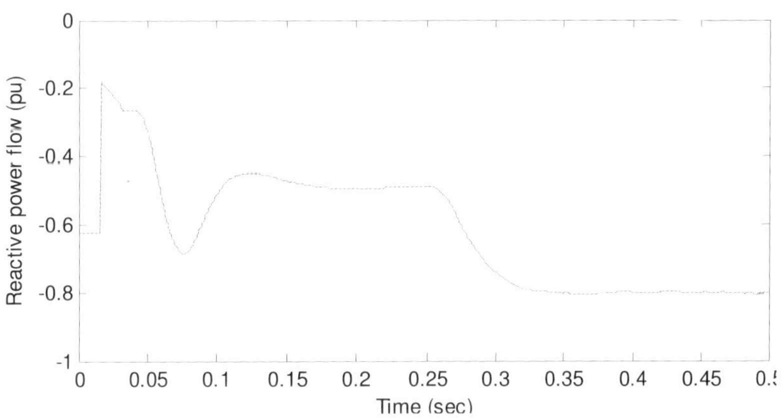

Figure: Reactive power flow through line 2

The transient stability of the system is analysed with the application of a three phase to ground fault in line 2. The real and reactive power references are fixed at 11 pu and -0.67 pu respectively. Fault is applied at 0.2 sec and cleared at 0.3 sec and the results are analysed. Figures below show the waveforms of the real and reactive power flow through line 2.

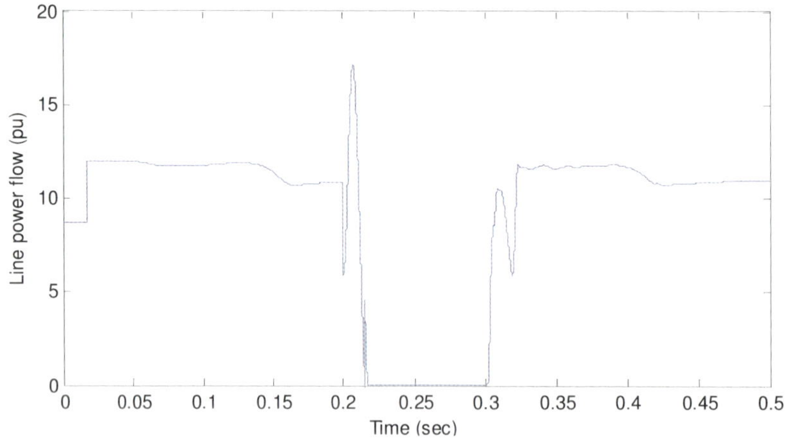

Figure: Real power flow through line 2

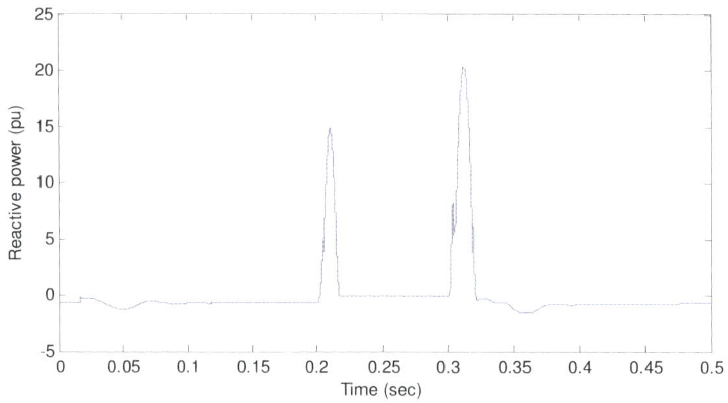

Figure: Reactive power flow through line 2

The injected voltage by SSSC, the current through line 2 and the voltage across the dc capacitor are shown in three figures below respectively. In the integrated system, the settling of the dc capacitor voltage is fast compared to the system with SSSC.

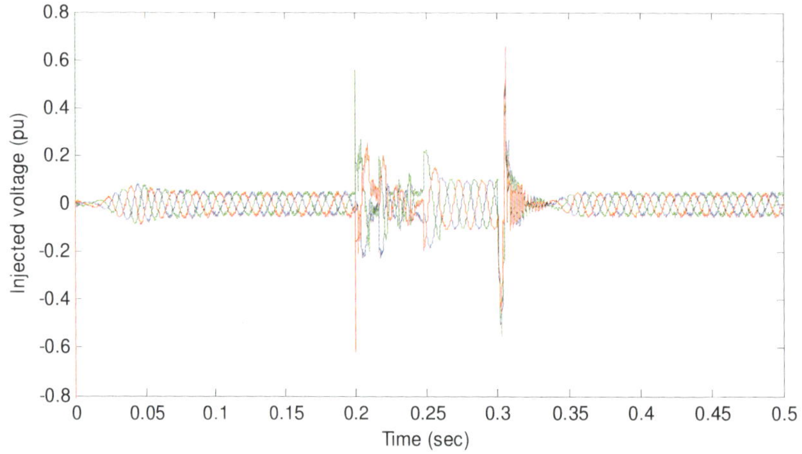

Figure: Injected voltage from SSSC

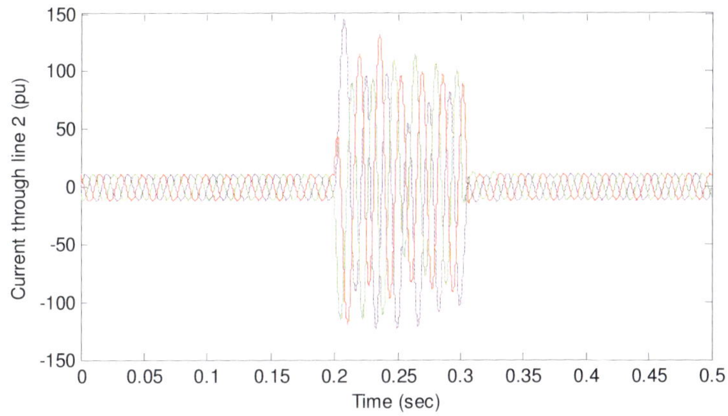

Figure: Current through line 2

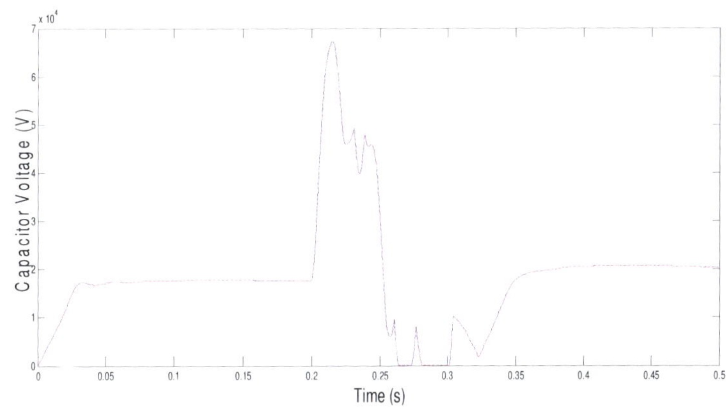

Figure: Voltage Across the DC Capacitor

Figures below show the waveforms of the electrical power output and load angle for machine. From the figures, it is clear that the oscillations are reduced for the system with SSSC/SMES combination. The overshoot and settling time are also less in the second case. Table below shows the comparative performances of the system with SSSC and with SSSC/SMES combination.

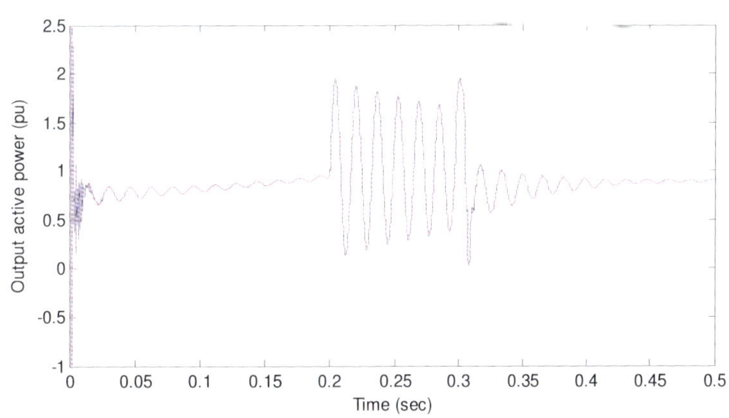

Figure: Electrical power output from generator 1

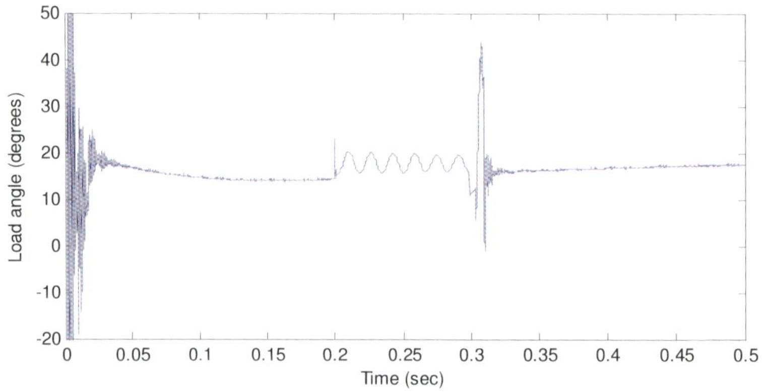

Figure: Load angle of machine 1

Table: Comparative results for the system with SSSC and with SSSC-SMES Combination

		With SSSC	SSSC-SMES COM BINATION
Real power in line 2	For Pref 8.5 pu	8.5	8.5
	For Pref 13 pu	12.8	13
Real power in line 2 (fault condition)	Peak value (pu)	20.2	17.1
Reactive power in line 2 (fault condition)	Peak value (pu)	24.1	20.4
Injected voltage	Final value (pu)	0.06	0.055
Electrical power output from generator 1	Peak value (pu)	2.14	1.94
Load angle in generator 1	Peak value (degrees)	51	44

Flywheel Energy Storage

Flywheel energy storage (FES) works by accelerating a rotor (flywheel) to a very high speed and maintaining the energy in the system as rotational energy. The energy is converted back by slowing down the flywheel.

Most FES systems use electricity to accelerate and decelerate the flywheel, but devices that directly use mechanical energy are being developed.

Advanced FES systems have rotors made of high strength carbon-composite filaments that spin at speeds from 20,000 to over 50,000 rpm in a vacuum enclosure and use magnetic bearings. Such flywheels can come up to speed in a matter of minutes—much quicker than some other forms of energy storage.

NASA G2 flywheel module

Main Components

A typical system consists of a rotor suspended by bearings inside a vacuum chamber to reduce friction, connected to a combination electric motor/electric generator.

Rotor

First generation flywheel energy storage systems use a large steel flywheel rotating on mechanical bearings. Newer systems use carbon-fiber composite rotors that have a higher tensile strength than steel and are an order of magnitude lighter.

Bearings

Magnetic bearings are necessary; in conventional mechanical bearings, friction is directly proportional to speed, and at such speeds, too much energy would be lost to friction.

The expense of refrigeration led to the early dismissal of low temperature superconductors for use in magnetic bearings. High-temperature superconductor (HTSC) bearings however may be economical and could possibly extend the time energy could be stored economically. Hybrid bearing systems are most likely to see use first. High-temperature superconductor bearings have historically had problems providing the lifting forces necessary for the larger designs, but can easily provide a stabilizing force. Therefore, in hybrid bearings, permanent magnets support the load and high-temperature superconductors are used to stabilize it. The reason superconductors can work well stabilizing the load is because they are good diamagnets. If the rotor tries to drift off center, a restoring force due to flux pinning restores it. This is known as the magnetic stiffness of the bearing. Rotational axis vibration can occur due to low stiffness and damping, which are inherent problems of superconducting magnets, preventing the use of completely superconducting magnetic bearings for flywheel applications.

Since flux pinning is the important factor for providing the stabilizing and lifting force, the HTSC can be made much more easily for FES than for other uses. HTSC powders can be formed into arbitrary shapes so long as flux pinning is strong. An ongoing challenge that has to be overcome before superconductors can provide the full lifting force for an FES system is finding a way to suppress the decrease of levitation force and the gradual fall of rotor during operation caused by the flux creep of SC material.

Parasitic losses such as friction, hysteresis, and eddy currents of both magnetic and conventional bearings in addition to refrigerant costs can limit the economical energy storage time for flywheels. However, further improvements in superconductors may help eliminate eddy current losses in existing magnetic bearing designs as well as raise overall operating temperatures. Even without such improvements, however, modern flywheels can have a zero-load rundown time measurable in years. (The 'zero-load rundown time' measures how long it takes for the device to come to a standstill when it is not connected to any other devices.)

Physical Characteristics

Compared with other ways of storing electricity, FES systems have long lifetimes (lasting decades with little or no maintenance; full-cycle lifetimes quoted for flywheels range from in excess of 10^5, up to 10^7, cycles of use), high energy densities (~ 130 W•h/kg, or ~ 500 kJ/kg), and large maximum power outputs. The energy efficiency *(ratio of energy out per energy in)* of flywheels can be as high as 90 percent. Typical capacities range from 3 kWh to 133 kWh.Rapid charging of a system occurs in less than 15 minutes.

Applications

Transportation

In the 1950s flywheel-powered buses, known as gyrobuses, were used in Yverdon, Switzerland, and there is ongoing research to make flywheel systems that are smaller, lighter, cheaper, and have a greater capacity. It is hoped that flywheel systems can replace conventional chemical batteries for mobile applications, such as for electric vehicles. Proposed flywheel systems would eliminate many of the disadvantages of existing battery power systems, such as low capacity, long charge times, heavy weight, and short usable lifetimes. Flywheels may have been used in the experimental Chrysler Patriot, though that has been disputed.

Flywheel systems have also been used experimentally in small electric locomotives for shunting or switching, e.g. the Sentinel-Oerlikon Gyro Locomotive. Larger electric locomotives, e.g. British Rail Class 70, have sometimes been fitted with flywheel boosters to carry them over gaps in the third rail. Advanced flywheels, such as the 133 kW•h pack of the University of Texas at Austin, can take a train from a standing start up to cruising speed.

During the 1990s Rosen Motors developed a gas turbine-powered series hybrid automotive powertrain using a 55,000 rpm flywheel to provide bursts of acceleration which the small gas turbine engine could not provide. The flywheel also stored energy through regenerative braking. The flywheel was composed of a titanium hub with a carbon fiber cylinder and gimbal mounted to minimize adverse gyroscopic effects on vehicle handling. The prototype vehicle was successfully road tested in 1997 but was never mass produced.

The Parry People Mover is a railcar which is powered by a flywheel. It was trialed on Sundays for 12 months on the Stourbridge Town Branch Line in the West Midlands, England during 2006 and

2007, and will be introduced as a full service by the train operator London Midland in December 2008 once two units have been ordered.

Uninterruptible Power Supply

Flywheel power storage systems in current production have storage capacities comparable to batteries and faster discharge rates. They are mainly used to provide load leveling for large battery systems, such as an uninterruptible power supply for data centers.

Flywheel maintenance in general runs about one-half the cost of traditional battery UPS systems. The only maintenance is a basic annual preventive maintenance routine and replacing the bearings every three years, which takes about four hours.

Laboratories

A long-standing niche market for flywheel power systems are facilities where circuit-breakers and similar devices are tested: even a small household circuit-breaker may be rated to interrupt a current of 10,000 or more amperes, and larger units may have interrupting ratings of 100,000 or 1,000,000 amperes. Obviously the enormous transient loads produced by deliberately forcing such devices to demonstrate their ability to interrupt simulated short circuits would have unacceptable effects on the local grid if these tests were done directly off building power. So typically such a laboratory will have several large motor-generator sets, which can be spun-up to speed over some minutes; then the motor is disconnected before a circuit breaker is tested. Other similar applications are in tokamak and laser experiments, where very high currents are also used for very brief intervals.

Amusement Ride

The Incredible Hulk roller coaster at Universal's Islands of Adventure features a rapidly accelerating uphill launch as opposed to the typical gravity drop. This is achieved through powerful traction motors that throw the car up the track. To achieve the brief very high current required to accelerate a full coaster train to full speed uphill, the park utilizes several motor generator sets with large flywheels. Without these stored energy units, the park would have to invest in a new substation and risk browning-out the local energy grid every time the ride launches.

Pulse Power

Since FES can store and release energy quickly, they have found a niche providing pulsed power.

Motor Sports

The FIA has included the use of KERS as part of its Formula 1 2009 Sporting Regulations. Using a continuously variable transmission (CVT), energy is recovered from the drive train during braking and stored in a flywheel. This stored energy is then used during acceleration by altering the ratio of the CVT. In motor sports applications this energy is used to improve acceleration rather than reduce carbon dioxide emissions—although the same technology can be applied to road cars to improve fuel efficiency.

Automobile Club de l'Ouest, the organizer behind the annual 24 Hours of Le Mans event and the Le Mans Series, is currently "studying specific rules for LMP1 which will be equipped with a kinetic energy recovery system."

Frequency Regulation

Beacon Power plans to open an electrical power frequency regulation plant before 2009. Lower carbon emissions, faster response times and ability to buy power at off-peak hours are among some advantages of using flywheels instead of traditional sources of energy for peaking power plants.

Advantages and Disadvantages

Flywheels are not affected by temperature changes as are chemical rechargeable batteries, nor do they suffer from memory effect. They are also less potentially damaging to the environment, being made of largely inert or benign materials. Another advantage of flywheels is that by a simple measurement of the rotation speed it is possible to know the exact amount of energy stored. However, use of flywheel accumulators is currently hampered by the danger of explosive shattering of the massive wheel due to overload.

One of the primary limits to flywheel design is the tensile strength of the material used for the rotor. Generally speaking, the stronger the disc, the faster it may be spun, and the more energy the system can store. When the tensile strength of a flywheel is exceeded the flywheel will shatter, releasing all of its stored energy at once; this is commonly referred to as "flywheel explosion" since wheel fragments can reach kinetic energy comparable to that of a bullet. Consequently, traditional flywheel systems require strong containment vessels as a safety precaution, which increases the total mass of the device. Fortunately, composite materials tend to disintegrate quickly once broken, and so instead of large chunks of high-velocity shrapnel, one simply gets a containment vessel filled with red-hot sand. Still, many customers of modern flywheel power storage systems prefer to have them embedded in the ground to halt any material that might escape the containment vessel.

When used in vehicles, flywheels also act as gyroscopes, since their angular momentum is typically of a similar order of magnitude as the forces acting on the moving vehicle. This property may be detrimental to the vehicle's handling characteristics while turning. On the other hand, this property could be utilized to keep the car balanced so as to keep it from rolling over during sharp turns. Conversely, the effect can be almost completely removed by mounting the flywheel within an appropriately applied set of gimbals, where the angular momentum is conserved without affecting the vehicle. This doesn't avoid the complication of gimbal lock, and thus a compromise between the number of gimbals and the angular freedom is needed. A single gimbal, for instance, could free a car for the 360 degrees necessary for regular driving. However, for instance driving up-hill would require a new gimbal mechanism with a new degree of freedom. Two gimbals would theoretically solve this problem and never lock unless your car rolls (something one carefully placed gimbal would not let the car do).

An alternative solution to the problem is to have two joined flywheels spinning synchronously in opposite directions. They would have a total angular momentum of zero and no gyroscopic effect. A problem with this solution is that when the difference between the momentum of each flywheel is anything other than zero the housing of the two flywheels would exhibit torque. You would, in effect, have two heavy spinning gears. Both wheels must be maintained at the same speed to keep the angular velocity at zero. Strictly speaking, the two flywheels would exert a huge torqueing moment at the central point, trying to bend the axle. However, if the axle were sufficiently strong, no gyroscopic forces would have a net effect on the sealed container, so no torque would be noticed.

Battery Energy Storage System

In batteries, the energy is stored electrochemically. Batteries are the most cost effective energy storage technologies available. A battery system is made up of a set of low-voltage / power battery modules connected in parallel and series to achieve a desired electrical characteristic. Batteries are "charged" when they undergo an internal chemical reaction under a potential applied to the terminals. They deliver the absorbed energy or "discharge" when they reverse the chemical reaction. Key factors of batteries for storage applications are: high energy density, high energy capability, round trip efficiency, cycling capability, life span and initial cost.

A power conversion is required to interface the battery with an ac system since they store dc charge. Small, modular batteries with power electronic converters can provide four quadrant operation (bidirectional current flow and bidirectional voltage polarity) with rapid response. Increased energy storage densities, greater cycling capabilities, higher reliability and lower cost are the recent advances in battery technologies.

Battery energy storage systems have recently emerged as one of the more promising near term storage technologies for power applications, providing a wide range of power system applications such as area regulation, area protection, spinning reserve and power factor correction. Several BESS units have been designed and installed in existing systems for the purposes of load levelling, stabilising and load frequency control. The power system operation and control can be improved with the integration of battery energy storage with FACTS devices.

There are some operational inconveniences with the BESS. Batteries cannot operate at high power levels for long time periods due to the chemical kinetics involved. Rapid, deep discharges lead to early replacement of the batter, since heating resulting in this kind of operation reduces battery lifetime. There are also environmental concerns related to battery storage due to toxic gas generation during battery charge/discharge. The disposal of hazardous materials presents some battery disposal problems.

Battery Technologies

There are a large range of battery technologies that form the generic term Battery Energy Storage System. This includes:

- VRLA or valve regulated lead acid batteries: These are the traditional batteries used in Uninterruptible Power Supplies and in existence since 1934, their cost is relatively low in comparison to other battery technologies. Deep cycle VRLA batteries are the type most suited to energy storage to ensure that they have a long enough life span for this use.

- Lithium ion batteries: This is the type of battery that is commonly found in modern consumer electronics such as your smartphone. They have a high energy density with the potential for even higher capacities (lithium ion technology is constantly changing and improving). The main benefit with lithium ion batteries are that they can be charged and discharged more often that VRLA batteries with less deterioration in terms of storage capacity of longevity of product. The main downside is cost although this is coming down progressively.
- Vanadium flow batteries: This battery uses vanadium ions to store chemical potential energy. They are currently quite bulky in size but offer the ability to completely discharge without any adverse effects and a long life time (20+ years).
- Sodium nickel batteries: These batteries use molten salt as the electrolyte. This allows the batteries to withstand greater temperature extremes as the cells themselves operate at 270+ degrees Celsius and are insulated so that the external chassis is just above ambient temperature.
- Liquid metal batteries: This battery technology uses liquid metals as both the electrodes and the electrolyte. As there is a constant regeneration between cycles, the electrodes do not degrade with time. The cost of liquid metal is currently the main stumbling block.

Gasholder

Gasholders were introduced at the beginning of the nineteenth century for storage of locally produced coal gas, which was used for the public illumination and later for domestic use.

It is necessary to begin by drawing a precise distinction between the words gasometer and gasholder because they are often colloquially used in the wrong sense. The suffix '-meter' stands for an instrument for measurement - a gasometer therefore is an instrument for the measurement of an amount of gas. A gasholder on the contrary is a functional building, which is used for the storage of locally manufactured gas.

Function

The gasholder was created by the necessity to store the local manufactured gas in a suitable recipient. This structure had to fulfill two basic requirements:

1. A variable capacity depending on the actual production and consumption of gas.
2. A gas tight construction to avoid gas emissions.

A simple principal met both of the requirements: the water-based system, which allows a changing volume and a water-sealed device. This system worked with two primary elements: a water tank and a lift for the gas. This lift was a kind of vessel, which was closed on the top and open on the bottom. The lift was immersed into the tank and rose and fell according to the actual content of the gas. The lift was also a kind of indicator or meter of the actual content. If the lift wasn't visible at all, the gasholder was empty. But if it reached the top, it was obviously full. The simple lift was often called bell reflecting its form.

The installation of a guide framing guaranteed the correct movement of the bell. Often some weights were put on top or on the inner bottom of the bell to reach the right pressure. These earlier gasholder forms were based on a low-pressure system: 75-125mm water column, respectively 7-12,5mmbar.

Thus the gasholder's first architectural form reflected its function.

Form and Structure

The first gasholders for public illumination had a small capacity because they served primarily for the main squares and streets of the cities. In the nineteenth century there was advantage of manufactured gas not only for public, but also for private use. The gasholders' shape and structure became biger and more significant in the cityscapes as a new kind of industrial architecture.

In the nineteenth und early twentieth century there were generally two different kinds of water sealed gasholders:

1. Bell-type gasholder (or single-lift gasholder), linear (column) or spiral-guided,
2. Telescopic gasholder (or multi-lift gasholder), linear (column-guided/ guide-framed) or spiral guided.

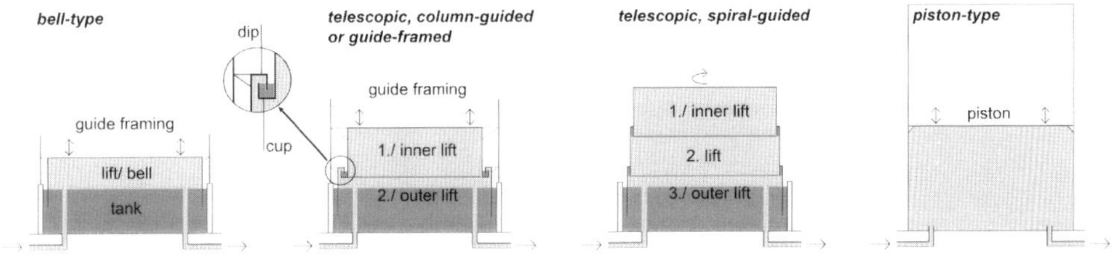

Figure: The principal development of the gasholder (from left to right): Bell-type gasholder – Telescopic gasholder, column-guided or guide-framed – Telescopic gasholder, spiral-guided – Piston gasholder

Bell-type Gasholder, Column-Guided

The first structural form of a gasholder was the bell-type gasholder (also single-lift gasholder). This gasholder was composed of a water tank, a single lift (bell) and a guide framing. First the gasholders had a rectangular ground plan, but soon were changed into a cylindrical form. The water tanks were built with wood, bricks, cast iron, steel or concrete. The position of the tank varied: from completely to partially underground or aboveground. Some designs also included banks, which surrounded the tank wall.

The guide framing ensured a linear movement for the lift and was usually designed as a riveted iron structure with columns of cast iron that were connected through horizontal framework girders at the top. Because of the appearance of the columns this kind of guiding was called column-guided. The former guide framings even had pillars of brick often in combination with a completely underground water tank.

The lift's structure was based on a cylindrical ring with a crown. Both were built with riveted single panel sheets on an interior substructure. With special wheels the lift ran along the vertical rails of the guide framing.

Today there are only a few original bell-type gasholders still remaining (e.g. in Salò). Later the spaces between the columns were also reinforced with wind bracing elements between the columns for stability (e.g. in Florence).

Figure: Florence, San Frediano: Bell-type gasholder, column-guided with an ornate ending "flame" on top of the columns, erected by: V-Ve-Moussy-Constructeur-Lyon, capacity: 8.000m³ . This was one of the smallest gasholders in Italy. This gasholder got a second so-called flying lift and augmented its capacity to 14.000m³

Telescopic Gasholder, Column-Guided/Guide-Framed

Because of the increasing demand for gas in the second half of the nineteenth century gasholders with a larger storage were required. In London a system was developed which added a second telescopic lift. So the original bell-type gasholder was upgraded to a two-lift-gasholder or telescopic gasholder that almost doubled its capacity. The most complex feature of this construction was the gas tight connection of the upper and the lower lift. This was resolved with an Ushaped cup (upstand) on the bottom edge of the inner lift filled with water for sealing. When the inner lift was full of gas its cup engaged with the down-turned cup (or dip) of the outer lift and pulled it up. Also the guide framing was extended with a second level, so that both lifts could run up and down linearly.

Figure: (a) (left): Rome, San Paolo: Telescopic/ 5-lift gasholder, guide-framed, patent of: August Klönne, Dortmund, capacity: 200.000m³. This in Rome called gazometro is one of the tallest remaining gasholders in Italy. (b) (right): Next to Klönne-type gasholder are three more telescopic gasholders.

In the early days of the telescopic gasholder, the column-guided structure was used for the guide framing. But the invention of the guide-framed design, an iron framework construction brought the advantage of a greater stability due to the inclusion of wind bracing elements and the possibility of prefabrication. Generally the guide framings number of horizontal levels stood for the number of lifts (that's not always valid for modern gasholders). This telescopic system was gradually increased to multi-lift gasholder with about seven lifts per gasholder.

Furthermore the telescopic concept allowed it to easily add another lift on existing gasholders (e.g. in Venice). This was a fast method to extend the storage capacity and also cheap since it did not require extending the guide framing. This so-called flying lift rose over the top of the guide framing and was only supported by the second lift (e.g. in Milan, Florence).

Another special architectural form of the telescopic gasholder had its own gasholder housing. This massive cylindrical facing formwork covered the real gasholder completely. Due to its often ornate and articulated façade (windows, cornice, pilaster, etc.) the gasholder housing seemed to belong rather to a sacred building than to a functional building. The advantages were the possibility of mounting the guide framing on this additional element and it protected the gasholder from abnormal weather and mechanical conditions (especially for the water sealing). The disadvantages were the increase of construction costs and the extended time of completion of construction.

Figure: Trieste, Broletto: Telescopic/ 2-lift gasholder, guide-framed with an architectonical ornate gasholder house, capacity: 20.000m^3

Today the majority of surviving gasholders in Italy are the multi-lift ones (e.g. in Milan, Turin, Venice, Treviso), but most of them are in a bad condition. Only a few gasholders have been repaired or even re-used with new functions like warehouses or parking lots (e.g. Rome).

Spiral-Guided Gasholder

At the end of the nineteenth century a new design changed the appearance of the gasholders - spiral-guided gasholder. This gasholder also worked with the water-sealed system and one or more lifts – but without an external guide framing. The lifts rose and fell spirally. For this movement the guide framing was mounted directly on the lifts: inclined guiding rails were fixed on the outer surface of the lift's cylinder. Each lift went in an opposite direction of the other. In order to maintain the lifts a staircase was added per lift, which rose and fell with its lift. There were different designs for the stairs, but the most popular ones were the stairs with a triangular fix structure.

The spiral-guided gasholders had a fascinating mechanism, which made the gasholder almost vanish into the ground when empty.

Today in Italy historic spiral-guided gasholders are rare (e.g. Milan, Vicenza – both demolished). This type was more widespread in the United Kingdom.

Figure: Milan-Bovisa: Telescopic/ 4-lift gasholder, spiral-guided, capacity: 130.000m^3

Piston Gasholder

At the end of the nineteenth century the water-sealed gasholders were built to such large dimensions, that the limit of construction was reached: the huge diameter and volume of the water tanks exceeded the possible ground pressure. Furthermore the water tank always had to be heated in the winter periods. To avoid these problems at the beginning of the twentieth century a new building technique was invented: the waterless piston gasholder. This new design appeared like a closed cylinder. Not only the shape, but also the elements of construction were different. This gasholder was composed of a closed cylindrical structure and an inserted piston that ran up and down according to the actual gas-content. Hidden in its cylinder the piston was not visible anymore than the lifts. Along the edge of the piston a sealing system was adapted which guaranteed a gas tight contact between the piston and the shell of the cylinder. Due to its dry sealing, the gasholder was also called dry or waterless gasholder.

The first piston gasholder was patented by the German company Maschinenfabrik Augsburg-Nürnberg (MAN) in 1914. The shell had a polygonal ground plan and was constructed from vertical support beams with straight panel plates in between and a roof. The piston was built with a frame structure. The fluid-sealing medium from the water-sealed gasholders was developed further to the piston oil-sealing system that ran along the inner side of the cylindrical shell. This oilfilm passed also the outer cup (dam) of the piston where a certain quantity of oil was always collected to ensure a positive gas seal. Both, the piston and the deck could be reached for maintenance by stairways and later also by elevators.

August Klönne from Dortmund invented the second patent for piston gasholders. Compared to the MAN-type the structural and technological details were different: the ground floor was not polygonal, but circular. Furthermore the sealing of the piston was not based on a fluid system. Klönne's sealing worked with a dry rubber sealing that had blocks filled with grease along the outer edge of the piston.

Today only a few piston gasholders are preserved (e.g. Bologna, Brescia, Genova). After being put out of service these tall gas-towers were seen a rather disturbing features of the cityscapes and have been removed to make way for other buildings (e.g. Florence, Marghera, Trieste).

Figure: Brescia: MAN Piston-type gasholder, capacity: 15.000m^3 - the previous identical gasholder was destroyed in the world war

Health and Safety

Gas holders are traditionally either above or below ground installations and the challenges can include the followings.

Cutting and Dismantling Operations

Changing the method of cutting from hot to cold eliminates the risk of sparks entering the holder. Oxy-fuel cutting (hot) uses fuel gases and oxygen to cut through the holder; changing the technique to cold can mean minimal burr production, fewer sparks and no dust. Using a specifically designed 'cold saw', the revolutions of the saw are slowed down so the blade chips the metal away rather than rotates at high speed.

Working in Confined Spaces

This involves the risks of gas build up, flooding and access issues. To overcome these challenges, atmospheric monitoring – ensuring site personnel aren't exposed to hazardous substances or significant flooding – is a must, with site personnel being equipped with personal gas monitors.

Cranes/Lifting Operations

Load testing on the site ground should be carried out in order to ensure the stability of machinery is maintained.

Contaminated Land

The ground on sites where town gas was previously manufactured can be heavily contaminated with a substance commonly known as 'Blue Billy', spent lime that becomes fouled with sulphur and sometimes cyanogens, noticeable by its distinctive blue colour. These areas should be covered with protective membranes and cordoned off to prevent access during the demolition process and removed by a specialist contractor during land remediation.

Environmental

The dismantling of gas holders leaves a massive issue around land remediation and the need to fill large holes in the ground.

Dismantling below-ground holders leaves a huge hole in the ground, sometimes in residential areas where children could potentially access the site. Full security on these sites need to be utilised to avoid injury and utilising waste excavated material from other projects i.e. from road infrastructure or local streetworks can be used as a backfill. This is a key environmental win as the spoil would otherwise go to landfill.

Subgrade vessels may be compromised by disturbance during demolition or subsequent site activity, settling or penetrating piping which could have corroded or been cut without plugging. This can allow the contaminated waters or sludges to leak into the environment, causing both soil and groundwater contamination.

The key to environmentally friendly demolition practices revolve around the proper, licensed removal and disposal of liquids and sludges from the vessels.

Thermal Energy Storage

Thermal energy storage involves the storage of heat in one of three forms; Sensible heat, latent heat and thermo-chemical heat storage.

Sensible heat storage is the most common method and has been employed for hundreds of years as hot water tanks. Sensible heat storage simply means changing the temperature of storage medium. The storage medium is most commonly water but rock, sand, clay and earth can also all be used.

Latent heat energy storage involves the storage of energy in Phase-Change Materials (PCM's). Thermal energy is stored and released with changes in the materials phase. The most common phase change to exploit is the solid-liquid transition, as the liquid–gas transition is impractical and solid-solid (crystalline structure) transitions usually have too low an energy density to be useful. When a PCM is heated initially it behaves like sensible heat energy storage and the materials temperature is increased. However, once the transition temperature is reached the material will continue to absorb heat at a constant temperature while it changes state. This heat absorbed at constant temperature is known as the latent heat of the transition. To retrieve the energy the PCM can be changed back from the liquid to the solid phase and the energy stored as latent heat is released.

Thermochemical heat energy storage involves storing heat energy in chemical bonds. A reversible chemical reaction which absorbs heat is used to absorb the heat energy that is to be stored. This reaction can then be reversed to release the stored heat. The most common reactions used for this process is the hydration of salts. The energy storage is based on the release of the heat of hydration. Hence, a salt hydrate storage system is charged by the endothermic thermal dehydration of the respective higher hydrated salt.

Thermal Energy Storage Technologies

A more specific categorization of thermal energy storage technologies is as shown below.

Ice Harvesting

For this technology, the evaporators arc placed over the TES Tank. The water return flow goes to the evaporators and ice blocks are formed. Afterwards there is a defrosting process in which the ice blocks are dropped into the tank where there is a mix of water and ice. This system may also be referred to as an ice harvester chiller because it is technically a water chiller and an icemaker.

Ice is formed during the charging cycle. Ice thickness can be as big as 10 mm, but this depends on cycle length.

Figure: Ice harvesting technology

External Melt Ice

This system builds ice and stores it on the exterior surface of a coil. The heat exchange coil used in this case is submerged in a tank full of water. A secondary coolant or liquid refrigerant passes through the heat exchanger coils and causes ice formation on the outside surface. During the discharge phase, the water is circulated from the top of the tank to the bottom of the tank. This water gets cold by melting the ice formation from their external surface.

Tanks used in this technology can be of several shapes. The most common shapes are vertical cylinders and rectangular parallelepiped cylinders. Ice annular thickness can vary a lot, but most ratings average at 36 mm.

Even formation is very important for the external ice melt process. That is why air is passed through the water both at the beginning of the charging cycle and at discharging cycle.

Benefits of this system include:

- No ice water flow restrictions, so ice melt rate can vary immensely.
- The system configuration is simpler than with other types of ice tanks.

Owing to these characteristics, internal melt ice systems are common in industrial and commercial applications.

Internal Melt Ice

Like in the external melt ice system, internal melt ice systems form ice on submerged tubes or pipes. This system needs a secondary coolant, mostly 75% water and 25% ethylene glycol, as the heat transfer fluid.

To discharge cooling, warm coolant circulates through the pipes and melts the ice. During the charging cycle, a standard chiller cools the glycol solution. As the cool solution (about -3°C) flows through the tubes within the tank, it causes ice to form on the outer surface.

Some advantages of this system include:

- Water does not leave storage container
- Simple controls-closed loop system.

Chilled Water

A chilled water thermal energy storage technology relies on the sensible heat characteristic of water. Temperature differential is very critical for this system because it determines tank volume.

TES comprises of two operational phases namely charging and discharging. During the charging phase, chilled water is entering at the bottom of the tank while return water is exiting at the top of the tank. During charging, the warm water is floating on the chilled water. Withdrawal of the water is at low velocity and in horizontal motion to maintain buoyancy as much as possible.

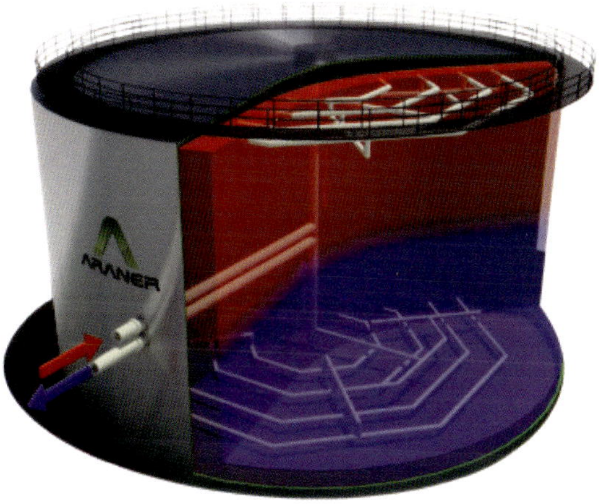

The separation between warm return water and stored cold water affects tank volume. Considering performance and cost factors, natural stratification emerges as the best alternative. Special diffusers maintain the movement of water into and out of the storage tank.

Multi Tank

A multi-tank design is essentially in the chilled water technology category, but it comprises of several tanks. Instead of a decoupler or bypass pipe in the single tank scenario, this system comprises of a number of tanks set in parallel from the chiller supply water temperature (for example 5.56°C) and water coming from the cooling consumers (for example at 14.44°C).

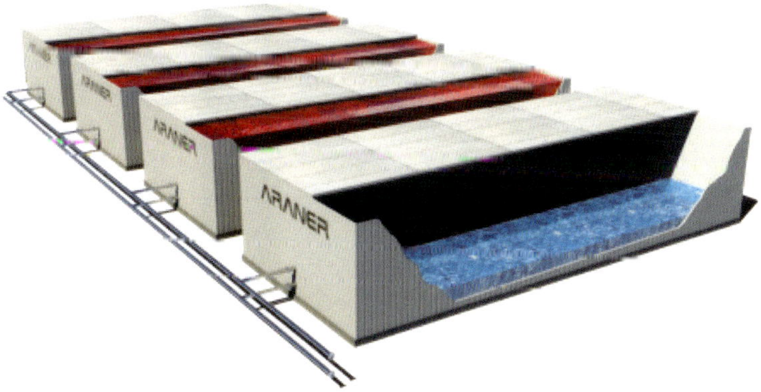

Strengths and Applications

Thermal energy storage – also called heat storage – has the main advantage that it does not involve any chemical conversion; hence it is typically simple, highly efficient and has a high cycle life. Yet, the economics and efficiency depend on the materials used and especially the integration of the storage into the overall system. A whole range of storage technologies and materials exists – with temperatures from 20°C (water) to 1000°C (metals or rocks), scales from kWh to GWh, and including phase change ("latent heat") or not ("sensible heat").

Thermal energy storage is the storage technology of choice for non-dispatchable heat production (e.g. process waste heat, solar heat), or where thermal energy such as steam is converted to power, (e.g. in a turbine). Instead of first generating the power and then storing it for later use, it is often better to first store the heat and then generate the power when it is actually needed.

More concrete, thermal energy storage can be applied:

- In concentrating solar power plants to provide dispatchable power, also during the night
- In thermal power plants to operate them more flexibly and perform quicker load changes
- In combined heat and power plants to provide security of heat supply and to temporally decouple the heat and power generation
- In process industries to recover and utilize heat that is otherwise lost
- As a component in other energy storage applications such as for example Compressed Air Energy Storage (CAES).

Power-to-Gas

The PtG process links the power grid with the gas grid by converting surplus power into a grid compatible gas via a two-step process: H_2 production by water electrolysis and H_2 conversion with an external CO or CO_2 source to CH_4 via methanation. The resulting CH4, known as substitute natural gas (SNG), can be injected into the existing gas distribution grid or gas storages, used as CNG motor fuel, or it can easily be utilised in all other well-established natural gas facilities. The total world storage capacity of natural gas is > 3600 TWh (total world power production from wind and solar power in 2012: 639 TWh).

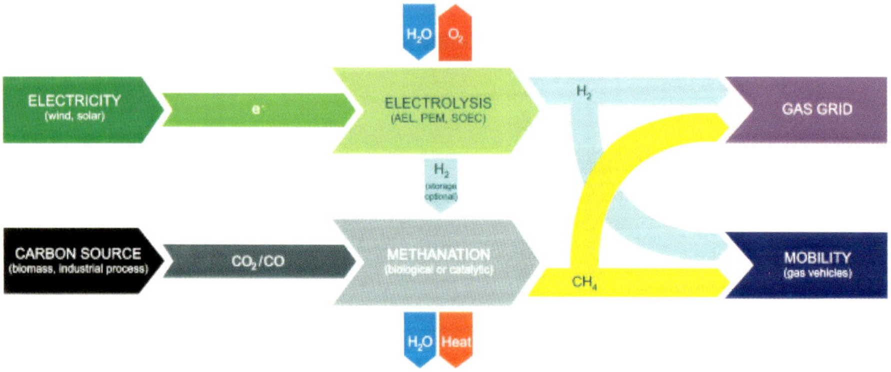

Figure: Exemplary Power-to-Gas process chain

The importance of PtG for handling high shares of renewable energies is being discussed at length. An alternative would be direct injection of H_2 into the gas grid. However, the amount of H_2 in the gas grid is limited by country specific standards and regulations to a maximum of 0–12 vol.%.

The main drawbacks of Power-to-Gas are a relatively low efficiency and high costs.

The Power-to-Gas process chain was first proposed in Japan in the 1980s–1990s. A global CO_2

recycling using sea water was proposed and a pilot plant in 2003 was built. A broader interest in Power-to-Gas has begun to grow (especially in Europe) in recent years driven by the increasing share of wind and solar power. In the meantime, there is a significant amount of Power-to-Gas research in different countries.

Five Power-to-Gas Parameters

Power-to-Gas has the potential to be a versatile, cross-sectoral technology supporting the implementation of the energy transition and integrating renewable energy into the different energy consumption areas. However, the current energy industry framework makes it impossible to use Power-to-Gas plants profitably. Five regulatory parameters need fine-tuning before Power-to-Gas is fully ready for market.

1. Hydrogen and methane made from renewable energy must be recognised as biofuels

The amended EU directives relating to the quality of petrol and diesel fuels and to the promotion of the use of energy from renewable energy sources must be transposed into German law without delay. This can be done via the Federal Emission Control Act. In this context, the inclusion in the list of biofuels of renewable fuels made from renewable power as opposed to biomass is an essential step.

2. Incentives designed to increase the flexibility of the electricity system

The remuneration system under the German Renewable Energy Sources Act (EEG) includes no incentives promoting the storage of electricity. The hardship regulation under the EEG hampers the development and use of storage solutions due to the fact that renewable electricity qualifies for remuneration whether or not it is actually used and fed into the grid. This regulation must be amended and an incentive for storage must be created.

3. Market launch of renewable hydrogen and methane by 2022 requires political support

The comprehensive launch of Power-to-Gas on a profitable basis must be accompanied by political measures. This includes the swift transposition into national law of the multiple inclusion of renewable gases in the fuel market.

4. Extension of tax reduction on natural gas as a fuel

The energy tax reduction on natural gas and automotive gas fuels, which is set to expire in 2018, must be extended. This objective should be included in the 2016 annual tax act.

5. Exemption of energy storage facilities from end user taxes

Power-to-Gas facilities are not end users and should not be categorised as such. Exemption will free them from inappropriate taxes and contributions associated with electricity purchases.

Storage Function

Power-to-gas systems may be deployed as adjuncts to wind parks or solar-electric generation. The excess power or off-peak power generated by wind generators or solar arrays may then be used at a later time for load balancing in the energy grid. Before switching to natural gas, the German gas networks were operated using towngas, which for 50–60 % consisted of hydrogen. The storage capacity of the German natural gas network is more than 200,000 GWh which is enough for several months of energy requirement. By comparison, the capacity of all German pumped storage power plants amounts to only about 40 GWh. The storage requirement in Germany is estimated at 16GW in 2023, 80GW in 2033 and 130GW in 2050. The transport of energy through a gas network is done with much less loss (<0.1%) than in a power network (8%). The storage costs per kilowatt hour are estimated at €0.10 for hydrogen and €0.15 for methane. The use of the existing natural gas pipelines for hydrogen was studied by the EU NaturalHy project and US DOE. The blending technology is also used in HCNG.

Efficiency

In 2013, the round-trip efficiency of power-to-gas-storage was well below 50%, with the hydrogen path being able to reach a maximum efficiency of ~ 43% and methane of ~ 39% by using combined-cycle powerplants. If cogeneration plants are used that produce both electricity and heat, efficiency can be above 60%, but is still less than pumped hydro or battery storage. However, there is potential to increase efficiency of power-to-gas storage. In 2015 a study published in Energy and Environmental Science found that by using reversible solid oxide electrochemical cells and recycling waste heat in the storage process a round-trip efficiency electricity to electricity of more than 70% can be reached at low cost.

Table: Overall energy conversion efficiency by pathway and fuel using electrolysis of water, plus methanation to produce methane

Fuel	Efficiency	Conditions
Pathway: Electricity→Gas		
Hydrogen	54–72 %	200 bar compression
Methane (SNG)	49–64 %	
Hydrogen	57–73 %	80 bar compression (Natural gas pipeline)
Methane (SNG)	50–64 %	
Hydrogen	64–77 %	without compression
Methane (SNG)	51–65 %	
Pathway: Electricity→Gas→Electricity		
Hydrogen	34–44 %	80 bar compression up to 60% back to electricity
Methane (SNG)	30–38 %	
Pathway: Electricity→Gas→Electricity & heat (cogeneration)		
Hydrogen	48–62 %	80 bar compression and electricity/heat for 40/45 %
Methane (SNG)	43–54 %	

Electrolysis Technology

Table: Advantages and disadvantages of the predominantly considered electrolysis technologies

Alkaline Electrolysis	
Advantage	Disadvantage
Commercial technology (high technology readiness level)	Limited cost reduction and efficiency improvement potential
Low investment electrolyser	High maintenance intensity
Large stack size	Modest reactivity, ramp rates and flexibility (minimal load 20%)
Extremely low hydrogen impurity (0.001%)	Stacks < 250 kW require unusual AC/DC converters
	Corrosive electrolyte deteriorates when not operating nominally

Proton Exchange Membrane Electrolysis (PEME)	
Advantage	Disadvantage
Reliable technology (no kinetics) and simple, compact design	High investment costs (noble metals, membrane)
Very fast response time	Limited lifetime of membranes
Cost reduction potential (modular design)	Requires high water purity

Solid Oxide Electrolysis Cell (SOEC)	
Advantage	Disadvantage
Highest electrolysis efficiency	Very low technology readiness level (proof of concept)
Low capital costs	Poor lifetime because of high temperature and affected material stability
Possibilities for integration with chemical methanation (heat recycling)	Limited flexibility; constant load required

Power to Hydrogen

In this method, electricity is used to split water into hydrogen and oxygen by means of electrolysis. The resulting hydrogen is injected into the natural gas grid or is used in transport or industry.

ITM Power won a tender in March 2013 for a Thüga Group project, to supply a 360 kW self-pressurising high pressure electrolysis rapid response PEM electrolyser *Rapid Response Electrolysis* Power-to-Gas energy storage plant. The unit produces 125 kg/day of hydrogen gas and incorporates AEG power electronics. It will be situated at a Mainova AG site in the Schielestraße, Frankfurt in the state of Hessen. The operational data will be shared by the whole Thüga group – the largest network of energy companies in Germany with around 100 municipal utility members. The project partners include: badenova AG & Co. kg, Erdgas Mittelsachsen GmbH, Energieversorgung Mittelrhein GmbH, erdgas schwaben GmbH, Gasversorgung Westerwald GmbH, Mainova Aktiengesellschaft, Stadtwerke Ansbach GmbH, Stadtwerke Bad Hersfeld GmbH, Thüga Energienetze GmbH, WEMAG AG, e-rp GmbH, ESWE Versorgungs AG with Thüga Aktiengesellschaft as project coordinator. Scientific partners will participate in the operational phase. It can produce 60 cubic metres of hydrogen per hour and feed 3,000 cubic metres of natural gas enriched with hydrogen into the grid per hour. An expansion of the pilot plant is planned from 2016, facilitating the full conversion of the hydrogen produced into methane to be directly injected into the natural gas grid.

Units like ITM Power's HGas generates hydrogen to be directly injected into the gas network as Power to Gas

In December 2013, ITM Power, Mainova, and NRM Netzdienste Rhein-Main GmbH began injecting hydrogen into the German gas distribution network using ITM Power HGas, which is a rapid response proton exchange membrane electrolyser plant. The power consumption of the electrolyser is 315 kilowatts. It produces about 60 cubic meters per hour of hydrogen and thus in one hour can feed 3,000 cubic meters of hydrogen-enriched natural gas into the network.

On August 28, 2013, E.ON Hanse, Solvicore, and Swissgas inaugurated a commercial power-to-gas unit in Falkenhagen, Germany. The unit, which has a capacity of two megawatts, can produce 360 cubic meters of hydrogen per hour. The plant uses wind power and Hydrogenics electrolysis equipment to transform water into hydrogen, which is then injected into the existing regional natural gas transmission system. Swissgas, which represents over 100 local natural gas utilities, is a partner in the project with a 20 percent capital stake and an agreement to purchase a portion of the gas produced. A second 800 kW power-to-gas project has been started in Hamburg/Reitbrook district and is expected to open in 2015.

In August 2013, a 140 MW wind park in Grapzow, Mecklenburg-Vorpommern owned by E.ON received an electrolyser. The hydrogen produced can be used in an internal combustion engine or can be injected into the local gas grid. The hydrogen compression and storage system stores up to 27 MWh of energy and increases the overall efficiency of the wind park by tapping into wind energy that otherwise would be wasted. The electrolyser produces 210 Nm^3/h of hydrogen and is operated by RH2-WKA.

In January 2011, Secure Supplies Hydrogen, an internationally certified engineering and warrantied power to gas vendor of quality Australian and USA made Hydrogen Fueling equipment, Hydrogen Fueled Engines and Hydrogen Home /Business Solar fuel cell Kits entered the market. Project's deploying (2015-2020) are focused on renewable power investments. Value adding Sites to produce a Hydrogen fuel and transfer storage gas. Secure Supplies value adds all solar wind, geo thermal and hydropower projects. Enabling owners to achieve a higher ROI, to produce hydrogen gas grids to fuel industry and a variety of applications emission free. End users include Port Operators, Farms, Green Community residential development or and Renewable power operators. Gas is sold to gas grid or transported and used to fuel fixed engines or fuel cell that run 24r to pump water or make power for business and communities projects. Hydrogen Equipment along with Engineering and Service provision makes Secure Supplies is well positioned to supply key markets globally.

The INGRID project started in 2013 in Apulia, Italy. It is a four-year project with 39 MWh storage and a 1.2 MW electrolyser for smart grid monitoring and control. The hydrogen is used for grid balancing, transport, industry, and injection into the gas network.

The surplus energy from the 12 MW Prenzlau Windpark in Brandenburg, Germany will be injected into the gas grid from 2014 on.

The 6 MW Energiepark Mainz from Stadtwerke Mainz, RheinMain University of Applied Sciences, Linde and Siemens in Mainz (Germany) will open in 2015.

Power to gas and other energy storage schemes to store and utilize renewable energy are part of Germany's Energiewende (energy transition program).

In France, the MINERVE demonstrator of AFUL Chantrerie (Federation of Local Utilities Association) aims to promote the development of energy solutions for the future with elected representatives, companies and more generally civil society. It also aims to experiment other reactors, catalysts. The synthetic methane produced by the MINERVE demonstrator (0.6 Nm3 / h of CH_4) is recovered as CNG fuel and as fuel in the boilers of the AFUL Chantrerie boiler plant. The installation was designed and built by the French SME Top Industrie, with the support of Leaf. It was received in November 2017 in the expected performances: 93.3% of CH_4. This project was supported by the ADEME and the ERDF-Pays de la Loire Region, as well as by several other partners: Conseil départemental de Loire -Atlantic, Engie-Cofely, GRDF, GRTGaz, Nantes-Metropolis, Sydela and Sydev.

Grid Injection Without Compression

The core of the system is a proton exchange membrane (PEM) electrolyser. The electrolyser converts electrical energy into chemical energy, which in turn facilitates the storage of electricity. A gas mixing plant ensures that the proportion of hydrogen in the natural gas stream does not exceed two per cent by volume, the technically permissible maximum value when a natural gas filling station is situated in the local distribution network. The electrolyser supplies the hydrogen-methane mixture at the same pressure as the gas distribution network, namely 3.5 bar.

Power to Methane

The power to methane method is to combine hydrogen from an electrolyzer with carbon dioxide and convert the two gases to methane using a methanation reaction such as the Sabatier reaction or biological methanation resulting in an extra energy conversion loss of 8%, the methane may then be fed into the natural gas grid if the purity requirement is reached.

ZSW (Center for Solar Energy and Hydrogen Research) and SolarFuel GmbH (now ETOGAS GmbH) realized a demonstration project with 250 kW electrical input power in Stuttgart, Germany. The plant was put into operation on October 30, 2012.

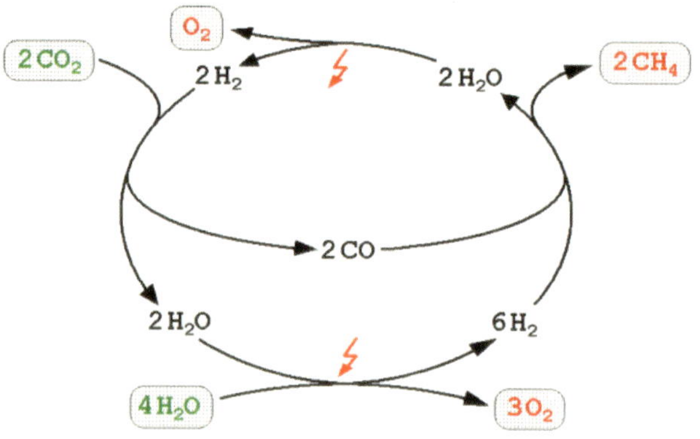

Methanation of CO_2 by electrolytically obtained hydrogen

The first industry-scale Power-to-Methane plant was realized by ETOGAS for Audi AG in Werlte, Germany. The plant with 6 MW electrical input power is using CO_2 from a waste-biogas plant and intermittent renewable power to produce synthetic natural gas (SNG) which is directly fed into the local gas grid (which is operated by EWE). The plant is part of the Audi e-fuels program. The produced synthetic natural gas, named Audi e-gas, enables CO_2-neutral mobility with standard CNG vehicles. Currently it is available to customers of Audi's first CNG car, the Audi A3 g-tron.

HELMETH Power-to-Gas Prototype

In April 2014 the European Union's co-financed and from the KIT coordinated HELMETH (Integrated High-Temperature ELectrolysis and METHanation for Effective Power to Gas Conversion) research project started. The objective of the project is the proof of concept of a highly efficient Power-to-Gas technology by thermally integrating high temperature electrolysis (SOEC technology) with CO_2-methanation. Through the thermal integration of exothermal methanation and steam generation for the high temperature steam electrolysis conversion efficiency > 85% (higher heating value of produced methane per used electrical energy) are theoretically possible. The process consists of a pressurized high-temperature steam electrolysis and a pressurized CO_2-methanation module. The project was completed in 2017 and achieved an efficiency of 76 % for the prototype with an indicated growth potential of 80 % for industrial scale plants. The operating conditions of the CO_2-methanation are a gas pressure of 10 - 30 bar, a SNG production of 1 - 5.4 m³/h (NTP) and a reactant conversion that produces SNG with H_2 < 2 vol.-% resp. CH_4 > 97 vol.-%. Thus, the generated substitute natural gas can be injected in the entire German natural gas network without limitations. As a cooling medium for the exothermic reaction boiling water is used at up to 300° C, which corresponds to a water vapour pressure of about 87 bar. The SOEC works with a pressure of up to 15 bar, steam conversions of up to 90% and generates one standard cubic meter of hydrogen from 3.37 kWh of electricity as feed for the methanation.

European demonstration project STORE&GO. The technological maturity of Power to Gas is evaluated in the European 27 partner project STORE&GO, which has started in March 2016 with a runtime of four years. Three different technological concepts are demonstrated in three different European countries (Falkenhagen/Germany, Solothurn/Switzerland, Troia/Italy). The technologies involved include biological and chemical methanation, direct capture of CO_2 from atmosphere, liquefaction of the synthesized methane to bio-LNG, and direct injection into the gas grid. The overall goal of the project is to assess those technologies and various usage paths under technical, economic, and legal aspects to identify business cases on the short and on the long term. The project is co-funded by the European Union's Horizon 2020 research and innovation programme (18 million euro) and the Swiss government (6 million euro), with another 4 million euro coming from participating industrial partners. The coordinator of the overall project is the research center of the DVGW located at the KIT.

Microbial Methanation

The biological methanation combines both processes, the electrolysis of water to form hydrogen and the subsequent CO_2 reduction to methane using this hydrogen. During this process, methane forming microorganisms (methanogenic archaea or methanogens) release enzymes that reduce the overpotential of a non-catalytic electrode (the cathode) so that it can produce hydrogen. This microbial power-to-gas reaction occurs at ambient conditions, i.e. room temperature and pH 7, at efficiencies that routinely reach 80-100%. However, methane is formed more slowly than in the Sabatier reaction due to the lower temperatures. A direct conversion of CO_2 to methane has also been postulated, circumventing the need for hydrogen production. Microorganisms involved in the microbial power-to-gas reaction are typically members of the order *Methanobacteriales*. Genera that were shown to catalyze this reaction are *Methanobacterium*, *Methanobrevibacter*, and *Methanothermobacter* (thermophile).

Power to Food

The synthetic methane generated from electricity can also be used for generating protein rich feed for cattle, poultry and fish economically by cultivating *Methylococcus capsulatus* bacteria culture with tiny land and water foot print. The carbon dioxide gas produced as by product from these plants can be recycled in the generation of synthetic methane (SNG). Similarly, oxygen gas produced as by product from the electrolysis of water and the methanation process can be consumed in the cultivation of bacteria culture. With these integrated plants, the abundant renewable solar / wind power potential can be converted in to high value food products with out any water pollution or green house gas (GHG) emissions.

Biogas-upgrading to Biomethane

In the third method the carbon dioxide in the output of a wood gas generator or a biogas plant after the biogas upgrader is mixed with the produced hydrogen from the electrolyzer to produce methane. The free heat coming from the electrolyzer is used to cut heating costs in the biogas plant. The impurities carbon dioxide, water, hydrogen sulfide, and particulates must be removed from the biogas if the gas is used for pipeline storage to prevent damage.

2014-Avedøre wastewater Services in Avedøre, Kopenhagen (Denmark) is adding a 1 MW electrolyzer plant to upgrade the anaerobic digestion biogas from sewage sludge. The produced hydrogen is used with the carbon dioxide from the biogas in a Sabatier reaction to produce methane. Electrochaea is testing another project outside P2G BioCat with biocatalytic methanation. The company uses an adapted strain of the thermophilic methanogen Methanothermobacter thermautotrophicus and has demonstrated its technology at laboratory-scale in an industrial environment. A pre-commercial demonstration project with a 10,000-liter reactor vessel was executed between January and November 2013 in Foulum, Denmark.

In 2016 Torrgas, Siemens, Stedin, Gasunie, A.Hak, Hanzehogeschool/EnTranCe and Energy Valley intend to open a 12 MW Power to Gas facility in Delfzijl (The Netherlands) where biogas from Torrgas (biocoal) will be upgraded with hydrogen from electrolysis and delivered to nearby industrial consumers.

Power to Syngas

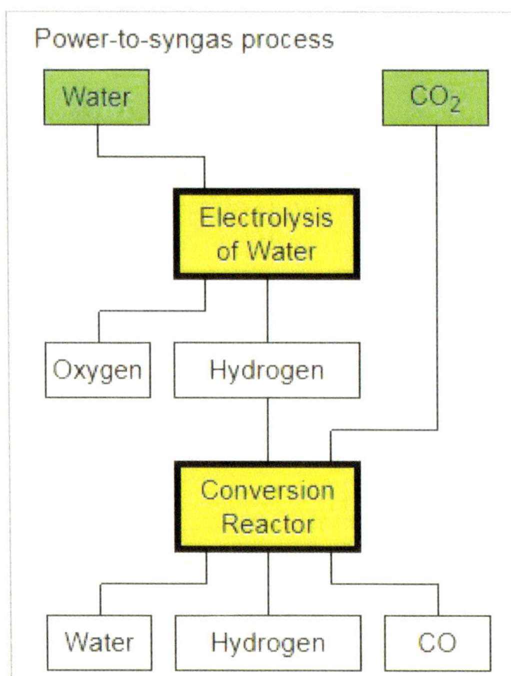

- 1st step: Electrolysis of Water (SOEC) –water is split into hydrogen and oxygen.
- 2nd step: Conversion Reactor (RWGSR) –hydrogen and carbon dioxide are inputs to the Conversion Reactor that outputs hydrogen, carbon monoxide, and water.
- $3H_2 + CO_2 \rightarrow (2H_2 + CO)^{syngas} + H_2O$
- Syngas is used to produce synfuels.

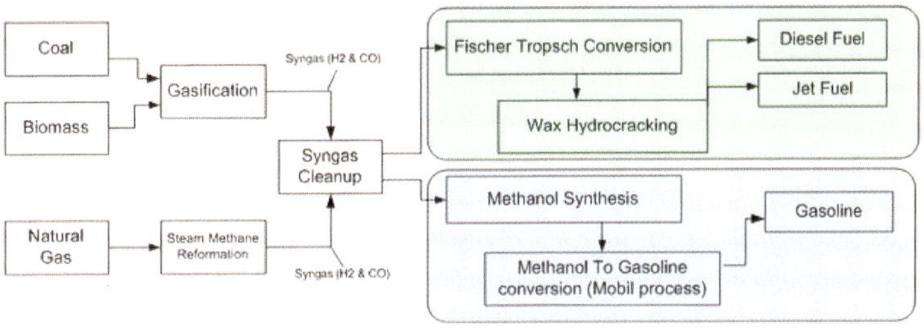

Power-to-syngasfeedstock is the same as feedstock derived from other sources.

Fuel Cell

A fuel cell is an electrochemical device that converts chemical energy into electricity. It is made up of two electrodes, each coated with a catalyst, separated by an electrolyte. During operation, the fuel cell consumes reactants, consisting of a fuel (such as hydrogen) and an oxidant (such as oxygen), and releases waste products (such as water and carbon dioxide). Electrons released by the fuel at one electrode travel though an external circuit to the other electrode, thereby producing electrical power. Fuel cells can operate virtually continuously as long as the necessary flows are maintained.

Methanol fuel cell. The actual fuel cell stack is the layered bi-cubic structure in the center of the figure

As power sources, fuel cells are useful for remote locations, such as remote weather stations, large parks, and rural areas. They are also valuable for spacecraft and certain military applications. A fuel cell system running on hydrogen can be compact and lightweight, with no major moving parts. In a recent application, known as micro combined heat and power, fuel cells are used to provide electricity for homes, offices, and factories, while the waste heat is used to produce hot air and water. In addition, excess power may be sold back to the grid.

Fuel Cell Design

In the archetypal hydrogen–oxygen proton exchange membrane fuel cell (PEMFC) design, a proton-conducting polymer membrane (the electrolyte) separates the anode and cathode sides. Each electrode is coated on one side with a catalyst that is typically a platinum group metal or alloy. This setup was called a "solid polymer electrolyte fuel cell" (SPEFC) in the early 1970s, before the proton exchange mechanism was well-understood.

In this cell, hydrogen diffuses to the anode catalyst where it dissociates into protons (hydrogen ions) and electrons. These protons often react with oxidants causing them to become what are commonly referred to as multi-facilitated proton membranes (MFPM). The protons are conducted through the membrane to the cathode, but the electrons (which are not conducted by the membrane) travel through an external circuit, supplying electric power. On the cathode catalyst, oxygen molecules react with the electrons (that have traveled through the external circuit) and protons to form water. In this example, the only waste product is water, in either the liquid or vapor form.

In addition to pure hydrogen, fuels used in fuel cells include hydrocarbons, methanol, and chemical hydrides. The waste products with these types of fuel are carbon dioxide and water. In addition to oxygen, oxidants used in fuel cells include chlorine and chlorine dioxide.

The electrode plates are usually made of metal (such as nickel) or carbon nanotubes, and are coated with a catalyst (such as platinum, nano iron powders, or palladium) for higher efficiency. Carbon paper separates them from the electrolyte. The electrolyte could be ceramic or an artificial membrane.

A typical PEM fuel cell produces a voltage from 0.6 V to 0.7 V at full rated load. Voltage decreases as current increases, due to several factors:

- Activation loss
- Ohmic loss (voltage drop due to resistance of the cell components and interconnects)
- Mass transport loss (depletion of reactants at catalyst sites under high loads, causing rapid loss of voltage).

To deliver the desired amount of energy, the fuel cells can be combined in series and parallel circuits, where series yield higher voltage, and parallel allows a stronger current to be drawn. Such a design is called a *fuel cell stack*. Further, the cell surface area can be increased, to allow stronger current from each cell.

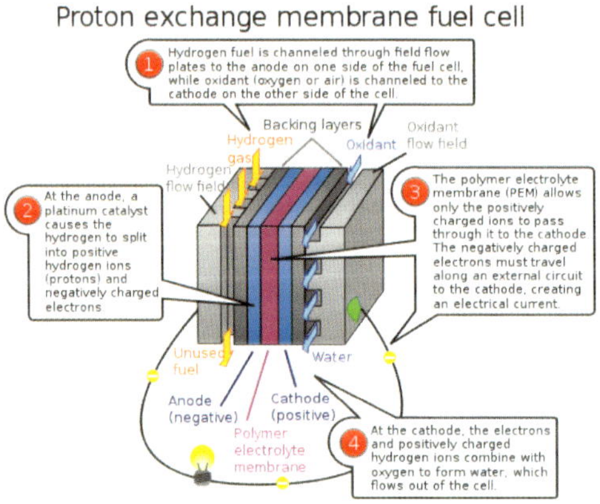

Figure: A low-temperature proton exchange membrane fuel cell (PEMFC). Components include: A bipolar plate as electrode, with in-milled gas channel structure, fabricated from conductive plastics (enhanced with carbon nanotubes for more conductivity); porous carbon papers; reactive layer, usually on the polymer membrane applied; polymer membrane

Condensation of water produced by a PEMFC on the air channel wall. The gold wire around the cell ensures the collection of electric current

Fuel Cell Design Issues

- Costs: In 2002, typical cells had a catalyst content of US$1000 per kilowatt of electric power output. In 2008 UTC Power has 400kw Fuel cells for $1,000,000 per 400kW installed costs. The goal is to reduce the cost in order to compete with current market technologies including gasoline internal combustion engines. Many companies are working on techniques to reduce cost in a variety of ways including reducing the amount of platinum needed in each individual cell. Ballard Power Systems have experiments with a catalyst enhanced with carbon silk which allows a 30 percent reduction (1 mg/cm^2 to 0.7 mg/cm^2) in platinum usage without reduction in performance. Monash University, Melbourne, uses PEDOT instead of platinum.

- The production costs of the PEM (proton exchange membrane): The Nafion membrane currently costs €400/m^2. The ITM Power membrane (a hydrocarbon polymer) is a new patent that may result in significant price reductions per square meter. in 2005 Ballard Power Systems announced that its fuel cells will use Solupor, a porous polyethylene film patented by DSM.

- Water and air management (in PEMFCs): In this type of fuel cell, the membrane must be hydrated, requiring water to be evaporated at precisely the same rate that it is produced. If water is evaporated too quickly, the membrane dries, resistance across it increases, and eventually it will crack, creating a gas "short circuit" where hydrogen and oxygen combine directly, generating heat that will damage the fuel cell. If the water is evaporated too slowly, the electrodes will flood, preventing the reactants from reaching the catalyst and stopping the reaction. Methods to manage water in cells are being developed like electroosmotic pumps focusing on flow control. Just as in a combustion engine, a steady ratio between the reactant and oxygen is necessary to keep the fuel cell operating efficiently.

- Temperature management: The same temperature must be maintained throughout the cell in order to prevent destruction of the cell through thermal loading. This is particularly challenging as the $2H_2 + O_2 \rightarrow 2H_2O$ reaction is highly exothermic, so a large quantity of heat is generated within the fuel cell.

- Durability, service life, and special requirements for some type of cells. Stationary applications typically require more than 40,000 hours of reliable operation at a temperature of -35°C to 40°C, while automotive fuel cells require a 5,000 hour lifespan (the equivalent of 150,000 miles) under extreme temperatures. Automotive engines must also be able to start reliably at -30°C and have a high power to volume ratio (typically 2.5 kW per liter).

- Limited carbon monoxide tolerance of the anode.

Difference from Electrochemical Batteries

Fuel cells do not store electrical energy but produce it as long as they are supplied with reactants, which must be replenished. By contrast, electrochemical batteries store electrical energy chemically, in a closed system. Additionally, the electrodes in a battery react and change as the battery is charged or discharged, but a fuel cell's electrodes are catalytic and relatively stable.

Types of Fuel Cells

Fuel Cell Name	Electrolyte	Qualified Power(W)	Working Temperature (°C)	Electrical efficiency	Status	Cost per Watt
Metal hydride fuel cell	Aqueous alkaline solution (e.g.potassium hydroxide)	?	above -20 (50% P_{peak} @ 0°C)	?	Commercial/Research	
Electro-galvanic fuel cell	Aqueous alkaline solution (e.g., potassium hydroxide)	?	under 40	?	Commercial/Research	
Direct formic acid fuel cell (DFAFC)	Polymer membrane (ionomer)	to 50 W	under 40	?	Commercial/Research	
Zinc-air battery	Aqueous alkaline solution (e.g., potassium hydroxide)	?	under 40	?	Mass production	
Microbial fuel cell	Polymer membrane or humic acid	?	under 40	?	Research	
Upflow microbial fuel cell (UMFC)		?	under 40	?	Research	
Reversible fuel cell	Polymer membrane (ionomer)	?	under 50	?	Commercial/Research	
Direct borohydride fuel cell	Aqueous alkaline solution (e.g., sodium hydroxide)	?	70	?	Commercial	
Alkaline fuel cell	Aqueous alkaline solution (e.g., potassium hydroxide)	10 kW to 100 kW	under 80	Cell: 60–70% System: 62%	Commercial/Research	
Direct methanol fuel cell	Polymer membrane (ionomer)	100 kW to 1 MW	90–120	Cell: 20–30% System: 10–20%	Commercial/Research	
Reformed methanol fuel cell	Polymer membrane (ionomer)	5 W to 100 kW	(Reformer)250–300 (PBI)125–200	Cell: 50–60% System: 25–40%	Commercial/Research	

Direct-ethanol fuel cell	Polymer membrane (ionomer)	up to 140 mW/cm2	above 25 ? 90–120	?	Research	
Direct formic acid fuel cell	Polymer membrane (ionomer)	?	25+	?	Research	
Proton exchange membrane fuel cell	Polymer membrane (ionomer) (e.g., Nafion or Polybenzimidazole fiber)	100 W to 500 kW	(Nafion)50–120 (PBI)125–220	Cell: 50–70% System: 30–50%	Commercial/Research	
RFC - Redox	Liquid electrolytes with redox shuttle & polymer membrane (Ionomer)	1 kW to 10 MW	?	?	Research	
Phosphoric acid fuel cell	Molten phosphoric acid (H_3PO_4)	up to 10 MW	150-200	Cell: 55% System: 40% Co-Gen: 90%	Commercial/Research	\$4-\$4.50 per watt
Molten carbonate fuel cell	Molten alkaline carbonate (e.g., sodium bicarbonate $NaHCO_3$)	100 MW	600-650	Cell: 55% System: 47%	Commercial/Research	
Tubular solid oxide fuel cell (TSOFC)	O^{2-}-conducting ceramic oxide (e.g., zirconium dioxide, ZrO_2)	up to 100 MW	850-1100	Cell: 60–65% System: 55–60%	Commercial/Research	
Protonic ceramic fuel cell	H^+-conducting ceramic oxide	?	700	?	Research	
Direct carbon fuel cell	Several different	?	700-850	Cell: 80% System: 70%	Commercial/Research	
Planar Solid oxide fuel cell	O^{2-}-conducting ceramic oxide (e.g., zirconium dioxide, ZrO_2 Lanthanum Nickel Oxide La_2XO_4,X= Ni,Co, Cu.)	up to 100 MW	850-1100	Cell: 60–65% System: 55–60%	Commercial/Research	

Fuel Cell Efficiency

The efficiency of a fuel cell is dependent on the amount of power drawn from it. Drawing more power means drawing more current, which increases the losses in the fuel cell. As a general rule, the more power (current) drawn, the lower the efficiency. Most losses manifest themselves as a voltage drop in the cell, so the efficiency of a cell is almost proportional to its voltage. For this reason, it is common to show graphs of voltage versus current (so-called polarization curves) for fuel cells. A typical cell running at 0.7 V has an efficiency of about 50 percent, meaning that 50 percent of the energy content of the hydrogen is converted into electrical energy; the remaining 50 percent will be converted into heat. (Depending on the fuel cell system design, some fuel might leave the system unreacted, constituting an additional loss.)

For a hydrogen cell operating at standard conditions with no reactant leaks, the efficiency is equal to the cell voltage divided by 1.48 V, based on the enthalpy, or heating value, of the reaction. For the same cell, the second law efficiency is equal to cell voltage divided by 1.23 V. (This voltage varies with fuel used, and quality and temperature of the cell.) The difference between these number represents the difference between the reaction's enthalpy and Gibbs free energy. This difference always appears as heat, along with any losses in electrical conversion efficiency.

Fuel cells do not operate on a thermal cycle. As such, they are not constrained, as combustion engines are, in the same way by thermodynamic limits, such as Carnot cycle efficiency. At times this is misrepresented by saying that fuel cells are exempt from the laws of thermodynamics, because most people think of thermodynamics in terms of combustion processes (enthalpy of formation). The laws of thermodynamics also hold for chemical processes (Gibbs free energy) like fuel cells, but the maximum theoretical efficiency is higher (83 percent efficient at 298K) than the Otto cycle thermal efficiency (60 percent for compression ratio of 10 and specific heat ratio of 1.4). Comparing limits imposed by thermodynamics is not a good predictor of practically achievable efficiencies. Also, if propulsion is the goal, electrical output of the fuel cell has to still be converted into mechanical power with the corresponding inefficiency. In reference to the exemption claim, the correct claim is that the "limitations imposed by the second law of thermodynamics on the operation of fuel cells are much less severe than the limitations imposed on conventional energy conversion systems." Consequently, they can have very high efficiencies in converting chemical energy to electrical energy, especially when they are operated at low power density, and using pure hydrogen and oxygen as reactants.

In Practice

For a fuel cell operated on air (rather than bottled oxygen), losses due to the air supply system must also be taken into account. This refers to the pressurization of the air and humidifying it. This reduces the efficiency significantly and brings it near to that of a compression ignition engine. Furthermore fuel cell efficiency decreases as load increases.

The tank-to-wheel efficiency of a fuel cell vehicle is about 45 percent at low loads and shows average values of about 36 percent when a driving cycle like the NEDC (New European Driving Cycle) is used as test procedure. The comparable NEDC value for a Diesel vehicle is 22 percent.

It is also important to take losses due to fuel production, transportation, and storage into account. Fuel cell vehicles running on compressed hydrogen may have a power-plant-to-wheel efficiency of 22 percent if the hydrogen is stored as high-pressure gas, and 17 percent if it is stored as liquid hydrogen.

Fuel cells cannot store energy like a battery, but in some applications, such as stand-alone power plants based on discontinuous sources such as solar or wind power, they are combined with electrolyzers and storage systems to form an energy storage system. The overall efficiency (electricity to hydrogen and back to electricity) of such plants (known as round-trip efficiency) is between 30 and 50 percent, depending on conditions. While a much cheaper lead-acid battery might return about 90 percent, the electrolyzer/fuel cell system can store indefinite quantities of hydrogen, and is therefore better suited for long-term storage.

Solid-oxide fuel cells produce exothermic heat from the recombination of the oxygen and hydrogen. The ceramic can run as hot as 800 degrees Celsius. This heat can be captured and used to heat water in a micro combined heat and power (m-CHP) application. When the heat is captured, total efficiency can reach 80-90 percent. CHP units are being developed today for the European home market.

Fuel Cell Applications

Type 212 submarine with fuel cell propulsion of the German Navy in dock

Fuel cells are very useful as power sources in remote locations, such as spacecraft, remote weather stations, large parks, rural locations, and in certain military applications. A fuel cell system running on hydrogen can be compact and lightweight, and have no major moving parts. Because fuel cells have no moving parts and do not involve combustion, in ideal conditions they can achieve up to 99.9999% reliability. This equates to around one minute of down time in a two year period.

The world's first certified Fuel Cell Boat (HYDRA), in Leipzig/Germany

A new application is micro combined heat and power, which is cogeneration for family homes, office buildings and factories. This type of system generates constant electric power (selling excess power back to the grid when it is not consumed), and at the same time produces hot air and water from the waste heat. A lower fuel-to-electricity conversion efficiency is tolerated (typically 15-20 percent), because most of the energy not converted into electricity is utilized as heat. Some heat is lost with the exhaust gas just as in a normal furnace, so the combined heat and power efficiency is still lower than 100 percent, typically around 80 percent. In terms of exergy however, the process is inefficient, and one could do better by maximizing the electricity generated and then using the electricity to drive a heat pump. Phosphoric-acid fuel cells (PAFC) comprise the largest segment of existing CHP products worldwide and can provide combined efficiencies close to 90 percent (35-50 percent electric + remainder as thermal). Molten-carbonate fuel cells have also been installed in these applications, and solid-oxide fuel cell prototypes exist.

Since electrolyzer systems do not store fuel in themselves, but rather rely on external storage units, they can be successfully applied in large-scale energy storage, rural areas being one example. In this application, batteries would have to be largely oversized to meet the storage demand, but fuel cells only need a larger storage unit (typically cheaper than an electrochemical device).

One such pilot program is operating on Stuart Island in Washington State. There the Stuart Island Energy Initiative has built a complete, closed-loop system: Solar panels power an electrolyzer which makes hydrogen. The hydrogen is stored in a 500 gallon tank at 200 PSI, and runs a ReliOn fuel cell to provide full electric back-up to the off-the-grid residence

The world's first Fuel Cell Ship HYDRA used an AFC system with 6.5 kW net output.

Suggested Applications

- Base load power plants.
- Electric and hybrid vehicles.
- Auxiliary power.
- Off-grid power supply.
- Notebook computers for applications where AC charging may not be available for weeks at a time.
- Portable charging docks for small electronics (for example, a belt clip that charges your cell phone or PDA).
- Smartphones with high power consumption due to large displays and additional features like GPS might be equipped with micro fuel cells.

Hydrogen Transportation and Refueling

The first public hydrogen refueling station was opened in Reykjavík, Iceland in April 2003. This station serves three buses built by DaimlerChrysler that are in service in the public transport net of Reykjavík. The station produces the hydrogen it needs by itself, with an electrolyzing unit (produced by Norsk Hydro), and does not need refilling: all that enters is electricity and water. Royal Dutch Shell is also a partner in the project. The station has no roof, in order to allow any leaked hydrogen to escape to the atmosphere.

The GM 1966 Electrovan was the automotive industry's first attempt at an automobile powered by a hydrogen fuel cell. The Electrovan, which weighed more than twice as much as a normal van, could travel up to 70mph for 30 seconds.

Toyota FCHV PEM FC fuel cell vehicle

The 2001 Chrysler Natrium used its own on-board hydrogen processor. It produces hydrogen for the fuel cell by reacting sodium borohydride fuel with Borax, both of which Chrysler claimed were naturally occurring in great quantity in the United States. The hydrogen produces electric power in the fuel cell for near-silent operation and a range of 300 miles without impinging on passenger space. Chrysler also developed vehicles which separated hydrogen from gasoline in the vehicle, the purpose being to reduce emissions without relying on a nonexistent hydrogen infrastructure and to avoid large storage tanks.

In 2005, the British firm Intelligent Energy produced the first ever working hydrogen run motorcycle called the ENV (Emission Neutral Vehicle). The motorcycle holds enough fuel to run for four hours, and to travel 100 miles in an urban area, at a top speed of 50 miles per hour. Honda is also going to offer fuel-cell motorcycles.

A hydrogen fuel cell public bus accelerating at traffic lights in Perth, Western Australia

There are numerous prototype or production cars and buses based on fuel cell technology being researched or manufactured. Research is ongoing at a variety of motor car manufacturers. Honda has announced the release of a hydrogen vehicle in 2008.

Type 212 submarines use fuel cells to remain submerged for weeks without the need to surface.

Boeing researchers and industry partners throughout Europe are planning to conduct experimental flight tests in 2007 of a manned airplane powered only by a fuel cell and lightweight batteries. The Fuel Cell Demonstrator Airplane research project was completed recently and thorough systems integration testing is now under way in preparation for upcoming ground and flight testing. The Boeing demonstrator uses a Proton Exchange Membrane (PEM) fuel cell/lithium-ion battery hybrid system to power an electric motor, which is coupled to a conventional propeller.

Market Structure

Not all geographic markets are ready for SOFC powered m-CHP appliances. Currently, the regions that lead the race in Distributed Generation and deployment of fuel cell m-CHP units are the EU and Japan.

Hydrogen Economy

Electrochemical extraction of energy from hydrogen via fuel cells is an especially clean method of meeting power requirements, but not an efficient one, due to the necessity of adding large amounts of energy to either water or hydrocarbon fuels in order to produce the hydrogen. Additionally, during the extraction of hydrogen from hydrocarbons, carbon monoxide is released. Although this gas is artificially converted into carbon dioxide, such a method of extracting hydrogen remains environmentally injurious. It must however be noted that regarding the concept of the hydrogen vehicle, burning/combustion of hydrogen in an internal combustion engine (IC/ICE) is often confused with the electrochemical process of generating electricity via fuel cells (FC) in which there is no combustion (though there is a small byproduct of heat in the reaction). Both processes require the establishment of a hydrogen economy before they may be considered commercially viable, and even then, the aforementioned energy costs make a hydrogen economy of questionable environmental value. Hydrogen combustion is similar to petroleum combustion, and like petroleum combustion, still results in nitrogen oxides as a by-product of the combustion, which lead to smog. Hydrogen combustion, like that of petroleum, is limited by the Carnot efficiency, but is completely different from the hydrogen fuel cell's chemical conversion process of hydrogen to electricity and water without combustion. Hydrogen fuel cells emit only water during use, while producing carbon dioxide emissions during the majority of hydrogen production, which comes from natural gas. Direct methane or natural gas conversion (whether IC or FC) also generate carbon dioxide emissions, but direct hydrocarbon conversion in high-temperature fuel cells produces lower carbon dioxide emissions than either combustion of the same fuel (due to the higher efficiency of the fuel cell process compared to combustion), and also lower carbon dioxide emissions than hydrogen fuel cells, which use methane less efficiently than high-temperature fuel cells by first converting it to high-purity hydrogen by steam reforming. Although hydrogen can also be produced by electrolysis of water using renewable energy, at present less than 3% of hydrogen is produced in this way.

Hydrogen is an energy carrier, and not an energy source, because it is usually produced from other energy sources via petroleum combustion, wind power, or solar photovoltaic cells. Hydrogen may be produced from subsurface reservoirs of methane and natural gas by a combination of steam reforming with the water gas shift reaction, from coal by coal gasification, or from oil shale by oil shale gasification. Electrolysis, which requires electricity, and high-temperature electrolysis/thermochemical production, which requires high temperatures (ideal for nuclear reactors), are two primary methods for the extraction of hydrogen from water.

As of 2005, 49.7 percent of the electricity produced in the United States comes from coal, 19.3 percent comes from nuclear, 18.7 percent comes from natural gas, 6.5 percent from hydroelectricity, 3% from petroleum and the remaining 2.8% mostly coming from geothermal, solar and biomass. When hydrogen is produced through electrolysis, the energy comes from these sources. Though the fuel cell itself will only emit heat and water as waste, pollution is often caused when generating the electricity required to produce the hydrogen that the fuel cell uses as its power source (for example, when coal, oil, or natural gas-generated electricity is used). This will be the case unless the hydrogen is produced using electricity generated by hydroelectric, geothermal, solar, wind or other clean power sources (which may or may not include nuclear power, depending on one's attitude to the nuclear waste byproducts); hydrogen is only as clean as the energy sources used to produce it. A holistic approach has to take into consideration the impacts of an extended hydrogen scenario, including the production, the use and the disposal of infrastructure and energy converters.

Nowadays low temperature fuel cell stacks proton exchange membrane fuel cell (PEMFC), direct methanol fuel cell (DMFC) and phosphoric acid fuel cell (PAFC) make extensive use of catalysts. Impurities poison or foul the catalysts (reducing activity and efficiency), thus higher catalyst densities are required. Limited reserves of platinum quicken the synthesis of an inorganic complex very similar to the catalytic iron-sulfur core of bacterial hydrogenase to step in. Although platinum is seen by some as one of the major "showstoppers" to mass market fuel cell commercialization companies, most predictions of platinum running out and/or platinum prices soaring do not take into account effects of thrifting (reduction in catalyst loading) and recycling. Recent research at Brookhaven National Laboratory could lead to the replacement of platinum by a gold-palladium coating which may be less susceptible to poisoning and thereby improve fuel cell lifetime considerably. Current targets for a transport PEM fuel cells are 0.2 g/kW Pt—which is a factor of 5 decrease over current loadings—and recent comments from major original equipment manufacturers (OEMs) indicate that this is possible. Also it is fully anticipated that recycling of fuel cells components, including platinum, will kick-in. High-temperature fuel cells, including molten carbonate fuel cells (MCFC's) and solid oxide fuel cells (SOFC's), do not use platinum as catalysts, but instead use cheaper materials such as nickel and nickel oxide, which are considerably more abundant (for example, nickel is used in fairly large quantities in common stainless steel).

References

- B. E. Conway (1999). Electrochemical Supercapacitors: Scientific Fundamentals and Technological Applications. Berlin: Springer. ISBN 0306457369. Retrieved May 2, 2013.
- Thermal-energy-storage-technologies: araner.com, Retrieved 18 July 2018
- Belkin, Andrey; et., al. (2017). "Recovery of Alumina Nanocapacitors after High Voltage Breakdown". Sci. Rep. Bibcode:2017NatSR...7..932B. doi:10.1038/s41598-017-01007-9.
- Thermal-energy-storage: energystoragesense.com, Retrieved 30 March 2018
- Guilherme de Oliveira e Silva; Patrick Hendrick (September 15, 2016). "Lead-acid batteries coupled with photovoltaics for increased electricity self-sufficiency in households". Retrieved July 20, 2016.
- Health-safety-environmental-impacts-gas-holder: fgould.com, Retrieved 10 June2018

- Aifantis, Katerina E.; Hackney, Stephen A.; Kumar, R. Vasant (March 30, 2010). High Energy Density Lithium Batteries: Materials, Engineering, Applications. John Wiley & Sons. ISBN 978-3-527-63002-8.
- Flywheel-energy-storage: newworldencyclopedia.org, Retrieved 19 March 2018
- Hubler, Alfred W. (Jan–Feb 2009). "Digital Batteries". Complexity. Wiley Periodicals, Inc. 14 (3): 7–8. Bibcode:2009Cmplx..14c...7H. doi:10.1002/cplx.20275.
- Pumped-hydroelectric-storage-phs: energystoragesense.com, Retrieved 10 July 2018
- Debord, Matthew (May 1, 2015). "Elon Musk's big announcement: it's called 'Tesla Energy'". Business Insider. Retrieved June 11, 2015.
- Pumped-storage, encyclopedia: energyeducation.ca, Retrieved 15 May 2018
- Content, Thomas. Johnson Controls, UW Open Energy Storage Systems Test Lab In Madison, Milwaukee, Wisconsin: Milwaukee Journal Sentinel, May 5, 2014.
- Battery-energy-storage-systems-bess, knowledge-base: powercontinuity.co.uk, Retrieved 19 April 2018

5

Energy Management

The management of energy production and consumption via planning and operation is termed as energy management. The chapter closely examines the key concepts of energy management to provide an extensive understanding of the subject. Topics like facility management, industrial energy management, energy management challenge, etc. have been detailed herein.

Energy management is the art and science of optimum use of energy to maximise profits (minimise costs) and thereby improve the economic competiveness. The energy should be used efficiently, economically and optimally.

Energy management can also be defined as the science involving planning, directing, controlling the supply and consumption of energy to maximise productivity and comforts and minimise the energy costs and pollution with conscious, judicious and effective use of energy.

The energy management involves strategy, policy, organizational changes, energy audit, energy conservation measures, administrative actions, training and awareness programmes, monitoring of energy conservation efforts etc.

Energy management is an important management function of every organization (like production, finance, marketing, planning, and design). Energy organization must have a written energy management policy document and the top management must be committed to implement the energy policy. The energy objectives must be known to energy executive and supervisor. The energy must be monitored vis-a-vis the production.

Steps for Energy Management

Energy management involves the following basic steps:

1. Energy management as Policy and Commitment.

2. Selection of energy manager and defining his responsibilities;

   i. Energy planning,

   ii. Monitoring energy consumption,

   iii. Planning energy conservation,

   iv. Implementing energy conservation measures,

   v. Achieve energy conservation objectives.

3. Formulation energy strategy and energy conservation plan.

4. Bring awareness and involvement at various levels by means of training programmes, workshops, communication, in-house journals.

5. Introduce suggestion scheme and award scheme.

6. Appoint energy audit team/consultants.

7. Obtain report on energy conservation measures.

8. Establish practice of monitoring energy consumption and effectiveness of energy conservation measures.

9. Adopt new technology measures.

10. Adopt recycling of scrap, avoid wastage etc.

11. Carry out modifications, retrofitting or replacement of existing plant/machinery so as to save energy.

Objectives of Energy Management

1. Objectives of Supply Side:

To formulate energy strategies, plan energy supply on short term, mid-term and long term basis and to ensure adequate supply of various forms of secondary (usable) energy to various consumers in the allocated geographical zone with minimum cost and minimum environmental pollution, to regulate energy flow.

2. Objectives of End-user side:

To select optimum energy forms for consumption and to optimize energy consumption of each form of energy for reducing energy costs and for improving productivity, standard of living and environment.

In accordance with this generic objective, every end-user organisation should have an energy objective statement in written form as a management policy statement. This is an obligatory function for every organisation on supply side and demand side in individual and national interest.

Management of Energy Consumption

1. Metering your Energy Consumption and Collecting The Data

As a rule of thumb: the more data you can get, and the more detailed it is, the better.

The old school approach to energy-data collection is to manually read meters once a week or once a month. This is quite a chore, and weekly or monthly data isn't nearly as good the data that comes easily and automatically from the modern approach.

The modern approach to energy-data collection is to fit interval-metering systems that automatically measure and record energy consumption at short, regular intervals such as every 15-minutes or half hour.

Detailed interval energy consumption data makes it possible to see patterns of energy waste that it would be impossible to see otherwise. For example, there's simply no way that weekly or monthly meter readings can show you how much energy you're using at different times of the day, or on different days of the week. And seeing these patterns makes it much easier to find the routine waste in your building.

2. Finding and Quantifying Opportunities to Save Energy

The detailed meter data that you are collecting will be invaluable for helping you to find and quantify energy-saving opportunities..

The easiest and most cost-effective energy-saving opportunities typically require little or no capital investment.

For example, an unbelievable number of buildings have advanced control systems that could, and should, be controlling HVAC well, but, unbeknown to the facilities-management staff, are faulty or misconfigured, and consequently committing such sins as heating or cooling an empty building every night and every weekend.

(NB "HVAC" is just an industry acronym for Heating, Ventilation and Air Conditioning. It's a term that's more widely used in some countries than others.)

And one of the simplest ways to save a significant amount of energy is to encourage staff to switch equipment off at the end of each working day.

Looking at detailed interval energy data is the ideal way to find routine energy waste. You can check whether staff and timers are switching things off without having to patrol the building day and night, and, with a little detective work, you can usually figure out who or what is causing the energy wastage that you will inevitably find.

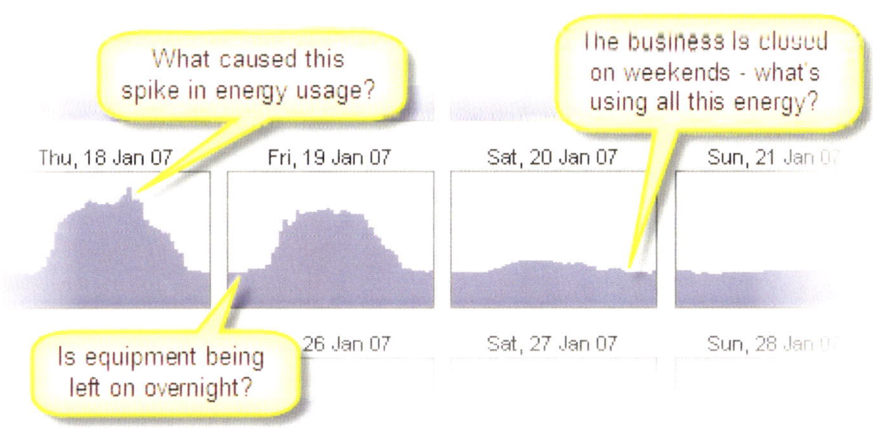

Detailed Energy Data is the Key to Finding the Easiest Energy Savings

And, using your detailed interval data, it's usually pretty easy to make reasonable estimates of how much energy is being wasted at different times. For example, if you've identified that a lot of energy is being wasted by equipment left on over the weekends, you can:

a. Use your interval data to calculate how much energy (in kWh) is being used each weekend.

b. Estimate the proportion of that energy that is being wasted (by equipment that should be switched off).

c. Using the figures from a and b, calculate an estimate of the total kWh that are wasted each weekend.

Alternatively, if you have no idea of the proportion of energy that is being wasted by equipment left on unnecessarily, you could:

i. Walk the building one evening to ensure that everything that should be switched off is switched off.

ii. Look back at the data for that evening to see how many kW were being used after you switched everything off.

iii. Subtract the target kW figure (ii) from the typical kW figure for weekends to estimate the potential savings in kW (power).

iv. Multiply the kW savings by the number of hours over the weekend to get the total potential kWh energy savings for a weekend.

Also, most buildings have open to them a variety of equipment- or building-fabric-related energy-saving opportunities, most of which require a more significant capital investment. You are probably aware of many of these, such as upgrading insulation or replacing lighting equipment, but good places to look for ideas include the Carbon Trust and Energy Star websites.

Although your detailed meter data won't necessarily help you to *find* these equipment- or building-fabric-related opportunites (e.g. it won't tell you that a more efficient type of lighting equipment exists), it will be useful for helping you to *quantify* the potential savings that each opportunity could bring. It's much more reliable to base your savings estimates on real metered data than on rules of thumb alone. And it's critically important to quantify the expected savings for any opportunity that you are considering investing a lot of time or money into – it's the only way you can figure out how to hone in on the biggest, easiest energy savings first.

3. Targeting The Opportunities to Save Energy

Just finding the opportunities to save energy won't help you to save energy - you have to take action to target them.

For those energy-saving opportunities that require you to motivate the people in your building, It can be hard work, but, if you can get the people on your side, you can make some seriously big energy savings without investing anything other than time.

As for those energy-saving opportunities that require you to upgrade equipment or insulation: assuming you've identified them, there's little more to be said. Just keep your fingers crossed that you make your anticipated savings, and be thankful that you don't work for the sort of organization that won't invest in anything with a payback period over 6 months.

Insulation - it usually works well, even when it looks like this..

4. Tracking your Progress at Saving Energy

Once you've taken action to save energy, it's important that you find out how effective your actions have been:

- Energy savings that come from behavioural changes (e.g. getting people to switch off their computers before going home) need ongoing attention to ensure that they remain effective and achieve their maximum potential.
- If you've invested money into new equipment, you'll probably want to *prove* that you've achieved the energy savings you predicted.
- If you've corrected faulty timers or control-equipment settings, you'll need to keep checking back to ensure that everything's still working as it should be. Simple things like a power cut can easily cause timers to revert back to factory settings - if you're not keeping an eye on your energy-consumption patterns you can easily miss such problems.
- If you've been given energy-saving targets from above, you'll need to provide evidence that you're meeting them, or at least making progress towards that goal.
- And occasionally you might need to prove that progress *isn't* being made (e.g. if you're at your wits' end trying to convince the decision makers to invest some *money* into your energy-management drive).

Energy Management in Operational Functions

Facility Management

Facility management is an important part of energy management, because a huge proportion (average 25 per cent) of complete operating costs are energy costs. According to the International Facility Management Association (IFMA), facility management is "a profession that encompasses multiple disciplines to ensure functionality of the built environment by integrating people, place, processes and technology."

The central task of energy management is to reduce costs for the provision of energy in buildings and facilities without compromising work processes. Especially the availability and service life of the equipment and the ease of use should remain the same. The German Facility Management Association (GEFMA e.V.) has published guidelines (e.g. GEFMA 124-1 and 124-2), which contain methods and ways of dealing with the integration of energy management in the context of a successful facility management. In this topic the facility manager has to deal with economic, ecological, risk-based and quality-based targets. He tries to minimize the total cost of the energy-related processes (supply, distribution and use).

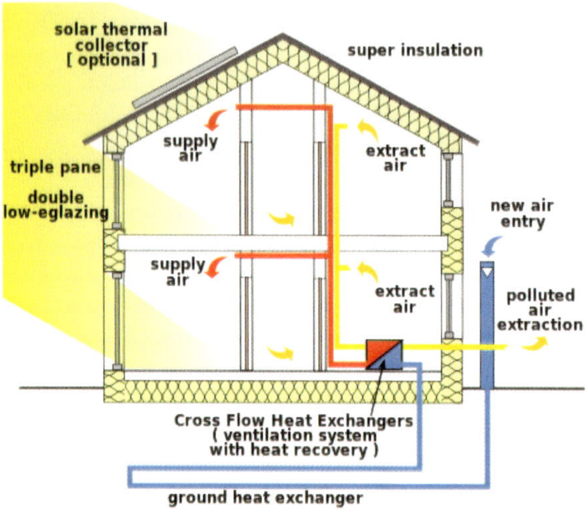

The Passivhaus uses a combination of low-energy building techniques and technologies.

The most important key figure in this context is kilowatt-hours per square meter per year (kWh/ m^2a). Based on this key figure properties can be classified according to their energy consumption.

- Europe: In Germany a low-energy house can have a maximum energy consumption of 70 kWh/m^2a.
- North America: In the United States, the ENERGY STAR program is the largest program defining low-energy homes. Homes earning ENERGY STAR certification use at least 15% less energy than standard new homes built to the International Residential Code, although homes typically achieve 20%-30% savings.

In comparison, the Passive house (Passivhaus in German) ultra-low-energy standard, currently undergoing adoption in some other European countries, has a maximum space heating requirement of 15 kWh/m^2a. A Passive House is a very well-insulated and virtually air-tight building. It does not require a conventional heating system. It is heated by solar gain and internal gains from people. Energy losses are minimized.

There are also buildings that produce more energy (for example by solar water heating or photovoltaic systems) over the course of a year than it imports from external sources. These buildings are called energy-plus-houses.

In addition, the work regulations manage competencies, roles and responsibilities. Because the systems also include risk factors (e.g., oil tanks, gas lines), you must ensure that all tasks are clearly described and distributed. A clear regulation can help to avoid liability risks.

Logistics

Carriage of Goods

Logistics is the management of the flow of resources between the point of origin and the point of destination in order to meet some requirements, for example of customers or corporations. Especially the core logistics task, transportation of the goods, can save costs and protect the environment through efficient energy management. The relevant factors are the choice of means of transportation, duration and length of transportation and cooperation with logistics service providers.

The logistics causes more than 14% percent of CO2 emissions worldwide. For this reason the term Green Logistics is becoming increasingly important.

Possible courses of action in terms of green logistics are:

- Shift to ecofriendly transport carrier such as railroad and waterway
- Route and load optimization
- Formation of corporate networks, which are connected by logistics service
- Optimizing physical logistics processes by providing a sophisticated IT support

Besides transportation of goods, the transport of persons should be an important part of the logistic strategy of organizations. In case of business trips it is important to attract attention to the choice and the proportionality of the means of transport. It should be balanced whether a physical presence is mandatory or a telephone or video conference is just as useful. Home Office is another possibility in which the company can protect the environment indirectly.

Energy Procurement

Procurement is the acquisition of goods or services. Energy prices fluctuate constantly, which can significantly affect the energy bill of organizations. Therefore poor energy procurement decisions can be expensive. Organizations can control and reduce energy costs by taking a proactive and efficient approach to buying energy. Even a change of the energy source can be a profitable and eco-friendly alternative.

Production

Production is the act of creating output, a good or service which has value and contributes to the utility of individuals. This central process may differ depending on the industry. Industrial companies have facilities that require a lot of energy. Service companies, in turn, do not need many materials, their energy-related focus is mainly facility management or Green IT. Therefore the energy-related focus has to be identified first, then evaluated and optimize.

Production Planning and Control

Usually, production is the area with the largest energy consumption within an organization. Therefore also the production planning and control becomes very important. It deals with the operational, temporal, quantitative and spatial planning, control and management of all processes that are necessary in the production of goods and commodities. The "production planner" should plan the production processes so that they operate in an energy efficient way. For example, strong power consumer can be moved into the night time. Peaks should be avoided for the benefit of a unified load profile.

The impending changes in the structure of energy production require an increasing demand for storage capacity. The Production planning and control has to deal with the problem of limited storability of energy. In principle there is the possibility to store energy electrically, mechanically or chemically. Another trend-setting technology is lithium-based electrochemical storage, which can be used in electric vehicles or as an option to control the power grid. The German Federal Ministry of Economics and Technology realized the significance of this topic and established an initiative with the aim to promote technological breakthroughs and support the rapid introduction of new energy storage.

Maintenance

Maintenance is the combination of all technical and administrative actions, including supervision actions, intended to retain an item in, or restore it to, a state in which it can perform a required function. Detailed maintenance is essential to support the energy management. Hereby power losses and cost increases can be avoided.

Energy Management Challenge

Through the energy efficiency it management is remain the key for the any industrial user across globe , to achieve the energy management goal for the federal government or industry the efficiency of water and energy resources play a vital role.

Examples of how it is possible to save energy and costs with the help of maintenance:

- Defrost the fridges

- Check the barometer of cars and trucks
- Insulation of hot systems
- Improve leaks in building envelopes

Energy Strategies

A long-term energy strategy should be part of the overall strategy of a company. This strategy may include the objective of increasing the use of renewable energies. Furthermore, criteria for decisions on energy investments, such as yield expectations, are determined. By formulating an energy strategy companies have the opportunity to avoid risks and to assure a competitive advance against their business rivals.

Potential Energy Strategies

According to Kals there are the following energy strategies:

- Passive Strategy: There is no systematic planning. The issue of energy and environmental management is not perceived as an independent field of action. The organization only deals with the most essential subjects.
- Strategy of short-term profit maximization: The management is concentrating exclusively on measures that have a relatively short payback period and a high return. Measures with low profitability are not considered.
- Strategy of long-term profit maximization: This strategy includes that you have a high knowledge of the energy price and technology development. The relevant measures (for example, heat exchangers or power stations) can have durations of several decades. Moreover, these measures can help to improve the image and increase the motivation of the employees.
- Realization of all financially attractive energy measures: This strategy has the goal to implement all measures that have a positive return on investment.
- Maximum strategy: For the climate protection one is willing to change even the object of the company.

In reality, you usually find hybrid forms of different strategies.

Energy Strategies of Companies

Many companies are trying to promote its image and time protect the climate through a proactive and public energy strategy. General Motors (GM) strategy is based on continuous improvement. Furthermore they have six principles: e.g. restoring and preserving the environment, reducing waste and pollutants, educating the public about environmental conservation, collaboration for the development of environmental laws and regulations.

Nokia created its first climate strategy in 2006. The strategy tries to evaluate the energy consumption and greenhouse gas emissions of products and operations and sets reduction targets accordingly. Furthermore, their environmental efforts is based on four key issues: substance management, energy efficiency, recycling, promoting environmental sustainability.

The energy strategy of Volkswagen (VW) is based on environmentally friendly products and a resource-efficient production according to the "Group Strategy 2018". Almost all locations of the Group are certified to the international standard ISO 14001 for environmental management systems.

When looking at the energy strategies of companies it is important to you have the topic greenwashing in mind. This is a form of propaganda in which green strategies are used to promote the opinion that an organization's aims are environmentally friendly.

Energy Strategies of Politics

Even many countries formulate energy strategies. The Swiss Federal Council decided in May 2011 to resign nuclear energy medium-dated. The nuclear power plants will be shut down at the end of life and will not be replaced. In Compensation they put the focus on energy efficiency, renewable energies, fossil energy sources and the development of water power.

The European Union has clear instructions for its members. The "20-20-20-targets" include, that the Member States have to reduce greenhouse gas emissions by 20% below 1990 levels, increase energy efficiency by 20% and achieve a 20% share of renewable energy in total energy consumption by 2020.

Ethical and Normative basis of the Energy Strategies

The basis of every energy strategy is the corporate culture and the related ethical standards applying in the company. Ethics, in the sense of business ethics, examines ethical principles and moral or ethical issues that arise in a business environment. Ethical standards can appear in company guidelines, energy and environmental policies or other documents.

The most relevant ethical ideas for the energy management are:

- Utilitarianism: This form of ethics has the maxim that the one acts are good or right, whose consequences are optimal for the welfare of all those affected by the action (principle of maximum happiness). In terms of energy management, the existence of external costs should be considered. They do not directly affect those who profit from the economic activity but non-participants like future generations. This error in the market mechanism can be solved by the internalization of external costs.
- Argumentation ethics: This fundamental ethical idea says that everyone who is affected by the decision, must be involved in decision making. This is done in a fair dialogue, the result is completely uncertain.
- Deontological ethics: The deontological ethics assigns individuals and organizations certain obligations. A general example is the golden rule: "One should treat others as one would like others to treat oneself." Therefore everyone should manage their duties and make an energy economic contribution.

Importance of Energy Management System

A range of energy management systems like SP1000 are being used widely. Your selection of the most appropriate energy management system would become easier if you know the importance and key features to look out in the system.

Reduce cost: With rocketing rates of power, such energy management equipment is high demand of the time. Make sure the system you select has enough arrangements and monitoring features to cut the cost by adequate power saving and fault detection.

Reduce risk: The selected system must be predictive to indicate opportunities of power saving by reducing the demand power and to reduce the risk of power shortage.

Reduce carbon emission: Apart from being able to conserve energy, the system must also be eco-friendly. Being eco-friendly is not only important to have a green image but it can also save further cost involving carbon taxes and managing carbon footprint.

Facility Management

Facility management is a term which is closely associated with building management. More broadly, facility management should not only be understood as general building management connected with everyday building operation but it should also include long term planning and focus on its users. This should already be essential in the preparatory phase of investment process focused on construction of building itself. Major part of operation costs and effectiveness of facility management processes is defined already at project of building. An important factor is also the ecological aspect of building influence on the living environment due to materials used for its construction as well as energy consumption necessary for its operation.

Theoretical and Methodological Aspects of the Issue

If we want to assess energy performance of buildings or its influence on environment, we have to analyse the whole building life cycle (starting with building materials up to the building deconstruction and its recycling) and not only the period of the building use. Such an analysis is called life cycle assessment. According to definition of Life Cycle Assessment by US Environmental Protection Agency Life cycle assessment is a technique for assessing the product influence on living environment connected with all life cycle phases (i.e. from raw materials extraction, material processing, manufacturing, distribution, use, repairs, maintenance, up to deconstruction or recycling). Life cycle assessment (LCA), is also known as life cycle analysis, ecological balance or from cradle-to-grave analysis). By the ISO 14040 (2006) LCA comes before the narrow view on environmental protection:

- Creating the list of corresponding energetic and material incomes and outcomes (emissions, waste) into living environment.
- Assessing possible influences (consequences) connected with the identified incomes and outcomes.
- Interpreting the results which enable erudite (qualified) decisions.

Facility management is defined as "a method whose task in organisations is to mutually harmonize employees, work activities and the work environment that includes principles of business administration, architecture and humanities and technical sciences". Facility management (FM) is an effective form of outreach business management which aims to provide relevant, cost-effective services to support the main business activities (core business) and allow them to optimize. At present, marked by the financial crisis, is the application of facility management most current because it provides savings and optimization of operating costs promotes increased employee performance and thus contributes to increase the profitability of the enterprise. The most common forms of application of facility management in the enterprise is a partial or complete outsourcing. According to the definition facility management is characterised by the interconnection of the three following areas:

- Area relating to employees, i.e. human resources and sociological aspects,
- Area of work activities, i.e. area of achievements and financing,
- Area of work environment, i.e. architecture and engineering.

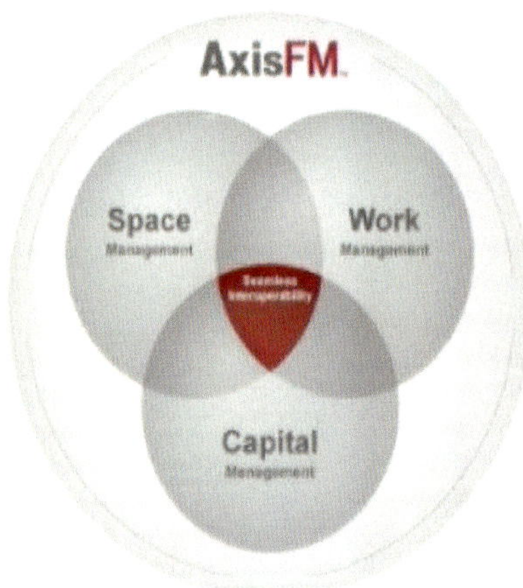

Figure: Axis of Facility Management

According to the definition, facility management is characterized by the interconnection of the three following areas - axis of facility management, the area relating to employees, i.e. human resources and sociological aspects, the area of work activities, i.e. area of achievements and financing and the area of work environment, i.e. architecture and engineering.

Energy Efficient Houses and their Life Cycle

With the continual increase of world population energy consumption is growing continually as well and in parallel there is an increase in CO_2 emissions. Buildings during their lifetime use approximately 40 % of energy, while its operation concerning heating and cooling counts for 60-70% of this consumed energy. The most suitable solution means building houses with low energy demands which do not harm the environment. Constructors can this way contribute to saving worldwide energy, lower emissions and what is crucial, they can considerably lower heating and cooling costs within the building. There exist several types of buildings which actively cope with the requirements of minimising the energy consumption of their operations. Tywoniak (2010) declarates that, globally these are understood as so called energy effective houses (EED), which include low energy houses (NED), energy passive houses (EPD), energy self-sufficient houses (END), energy independent -autarkic houses (EAD) as well as plus energy houses. In this article we will closely focus only on low energy and energy passive houses. Division of these buildings according to heat consumption is presented in figure below.

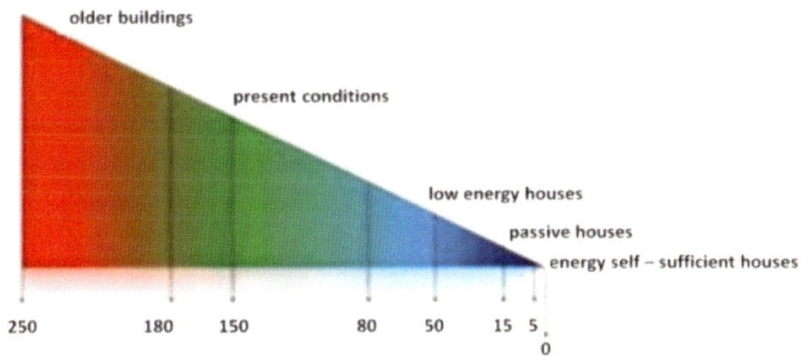

Figure: Division of buildings according to heat consumption (kWh/m2/per annum)

Life cycle of a building can be according to United Nations International Development Organization (UNIDO) as well as conditions of the Slovak Republic defined into three basic phases. Differences are presented in figure below:

	Building life cycle phases			
	Building acquisition		Building use	Building deconstruction
UNIDO	Pre-investment phase	Investment phase	Operational phase	
Slovak Republic	Phase of building preparation	Phase of building acquisition	Phase of building use	Phase of building deconstruction

Figure: Life cycle of building according to the UNIDO and conditions of the Slovak Republic

Phases of building life cycle of energy efficient buildings (acquisition, use, deconstruction), in general do not differ from classic buildings, but in more detailed specification there are certain differences in the phases of its acquisition and use.

In the phase of acquisition which includes initial study, preparation, design and project realisation leading to final construction the following criteria have to be fully respected:

- When locating the building into the terrain it is necessary to consider cardinal points and the character of the surrounding buildings with the main requirements of suitable big windows with the South orientation,
- Avoid creation of thermal bridges and secure accumulation of passive energy profits in winter and shading the building against overheating in summer,
- To avoid energy loses it is important to design air tight walls, choose high quality windows, roof construct,
- Controlled ventilation with heat recovery (ventilation with back trapping of heat) will secure necessary income of fresh air with the minimum energy loss.

The phase of using energy sufficient building has its advantages but of course some risks. The advantages in comparison with classic brick buildings are:

- saving energy costs for heating represent about 80 %,
- high quality of interior air climate,
- if the building is well insulated, in summer the interior does not get overheated, controlled ventilation can divert moisture and constant flow of fresh, clean and dust free air can circulate the building.

On the other hand a disadvantage is the increase of investments in comparison with the classic building when considering improved insulation of the building envelope (more and higher quality heat insulation with specific characteristics), investing into more quality windows, doors, heat recovery unit and requirements for highly qualified and quality construction details. These increased initial investments are though compensated by energy cost saving, which during a building life cycle represent a quick return on investment.

Cost of Building Life Cycle

Costs of building life cycle represent the total costs from all building life cycle phases, i.e. costs for acquisition, use and deconstruction.

It is obvious that costs of each building life cycle phase are influenced by various factors such as the purpose, space, construction and material solutions of the building design. Financial costs of each phase of the building life cycle cannot be mutually directly compared. This comes out from the change of money value over the time which cannot be compared during usually a long lasting building life cycle. Therefore it is necessary to calculate all financial values of regular periods to its present value within certain time period. This time period is usually the beginning of the building acquisition.

Method of building life cycle costs is a way of calculation of present cost value for its acquisition, use and deconstruction and serves for those building solution variations which require the lowest cost for its operation and running. The basic prerequisite of the use of this method is the fact that for the investor are, except present costs, important also cost for future ownership and use of the building. Costs of the life cycle should include at least 3 groups of costs, corresponding with real time “t“ within the overall considered time spam:

1. $C_{acq,t}$ = costs for acquisition of building, concerning the period t,
2. $C_{use,t}$ = costs for use of building, concerning the period t,
3. $C_{dec,t}$ = costs for deconstruction of building, concerning the period t,

Then to set the costs of building life cycle (Ct), concerning the period t the following pattern is valid:

$$Ct = C_{acq,t} + C_{use,t} + C_{dec,t}$$

These three groups can be according to their needs and availability of input data divided also into more detailed cost entries. Calculation of total costs of the building life cycle for i- variation of building design lies in addition of all building life cycle costs. Current values of all concerned costs are calculated within one given time t=1 with the pattern:

$$TLCC_i = \sum_{t=1}^{t=T} C_t x(1+u)(1+u)^{-t}$$

TLCC i = total building life cycle costs for variation i = 1,..., n

n = number of considered variations of building designs t = 1, 2,..., T

T = length of building life cycle in the number of individual periods, t

u = interest rate in absolute formulation,

Ct = building life cycle costs, concerning one particular period t

When energy use is deliberately monitored, controlled, and conserved, decreases in utility consumption and overall costs can be realized without sacrificing facilities operations. Such energy management techniques can take on many shapes and sizes. Following are strategies facility management executives can use to increase efficiency while overcoming potential costly challenges.

Actively manage real-time energy use: Proactive, real-time data management can expose a wide range of unknown challenges associated with occupancy, building use, and peaks in utility usage. For example, my firm, Southland Energy, installed a comprehensive metering system for a data center customer, monitoring everything from air and water flows, to very specific details of the data center. The real-time data allowed the building operators to identify potential issues instantaneously, implement corrective actions to prevent critical shutdowns, and manage loads before they affected the entire system.

Actively manage what is measureable: Use advanced metering and energy management systems (EMS) to capture real-time data, ensure its accuracy and, in turn, address specific issues. For example, a K-12 school installed an energy dashboard that managed the overall facility while actively engaging faculty and students. The customer could view how the systems were operating and how much they were saving based on their actions and system improvements.

In instances where building owners have utility monitoring equipment but no collection or processing software, the meters or monitoring equipment become stranded assets. This is because millions of data points have to be gathered and processed manually, multiple times during the year. A sophisticated metering system equipped with the proper EMS software will automatically collect, process, and format these data points in real time, if not hourly. The ability to process these useful data points into an easy to use format improves the overall system effectiveness and functionality.

Actively manage energy consumption. Use collected data to build a strategy that manages costs and consumption on a daily, weekly, monthly, and annual basis.. Load shifting and demand limiting systems were implemented to limit customer loads during peak hours and reduce costs.

Limiting peak demand consumption offers additional benefits that are not always easy to identify or claim. For example, during peak hours, utilities run "peaker plants" to meet demands from the grid. However, these plants are often older and less efficient electricity generation plants, with the sole purpose to run periodically to meet demand. Reducing peak demand during summer months saves electricity costs and overall greenhouse gas emissions per kW.

Managing consumption allows for early detection of improper set points, schedule misalignments, and equipment/system failures. Analyzing trends of metered points over days, weeks, months, and years helps to pinpoint irregularities, leaks, and excessive run times. The proper system can flag leaks, changes in occupancy, occupant set point changes, and energy and water waste.

Have a holistic plan. Without clear direction and an action plan, it is difficult to make a meaningful impact beyond the "low hanging fruit." A holistic plan is critical to leverage overall savings and provide a mix of improvements for substantial results. Facility leaders often benefit from a holistic plan that bundles low hanging fruit such as lighting and building automation measures with longer paybacks such as renewable energy. This evaluates all possible savings including water, waste, energy, and system/facility reliability to package the appropriate measures for the facility's goals and financial requirements.

Secure leadership buy-in and support. Real, holistic changes will not be attainable without direct involvement and support from leadership. It is critical to engage leadership and key decision makers that impact the financials of facility operations.

Negotiate. Negotiate supply contracts with third-party marketers to reduce energy costs. Southland Energy has worked with several customers on energy specific solutions and identified opportunities to help them negotiate utility rates efficiently. The firm worked with a condominium high-rise customer and was able to negotiate a 5% reduction of utility rates, while also evaluating improvement measures. These results can be achieved when we know historical consumption and have accurate projections for consumption; take advantage of available curtailment, arbitrage, and/or demand response programs; and issue pricing RFPs to multiple suppliers to ensure best market value.

Take risks. Major changes that drastically reduce consumption can require difficult steps and decisions, but reward outweighs risk if changes are properly managed. The ability of a trusted advisor to leverage in-house capabilities coupled with a strong commitment to change helps alleviate the burden.

Take action but don't "over study." While energy studies and audits are useful and provide direction, in most cases data is already available and can be used to take action sooner rather than delaying it. Develop a plan for action with available information.

Many facility executives complete several energy audits but do not act on a path forward. Southland Energy has performed such studies and audits, while also leveraging customers' existing studies and audits to implement their suggestions. From simple lighting retrofits to complicated central utility plants and renewable energy systems, customers have turned improvement ideas into a reality.

Partner with those who can supplement in-house knowledge. Relationship building and partnering are keys to building an effective energy management strategy. This can involve partnering with other facility management and energy management leaders; energy services companies (ESCOs); industry subject matter experts, such as the U.S. Department of Energy or consultants; utilities; and other knowledge centers, such as the Association of Energy Engineers (AEE).

Establish an occupant behavioral awareness program. Technology implementation and building retrofits are only part of the equation. Occupants have a big impact on a building's efficiency and investments made. Education is key to the behavioral process, and empowering occupants with knowledge and resources will help increase energy savings as they can realize the impact through efficiency or financial gain.

Industrial Energy Management

Organizations that successfully manage energy have business processes to plan, monitor, and control energy use, just as they do for other corporate priorities, such as labor, materials, and other costs. For these organizations, energy management is "business as usual".

Effective Features of Energy Efficiency Management Systems

Energy management made based on Plan-Do-Check Act model, which serve as an organizational culture of stable advancement in energy efficiency. The culture of stable advancement assures that the set of objectives are obtained in a progressive manner. It also assures that those set of objectives are practical, obtainable and accommodates the available resources of the firm.

Plan Phase

Planning has been defined as (1) outlining a course of action to achieve an objective, (2) devising a method to attain a defined goal, and (3) developing a specific way to obtain a desired result. Each definition contains two elements: Knowing what is wanted to be accomplished and deciding how to accomplish it. The energy planning process consists of three steps:

- deciding what is wanted to be accomplished, or setting objectives
- deciding how to accomplish the objectives, or outlining procedures
- deciding how to apply the skills needed to meet objectives, or assigning responsibilities. This step is necessary because the manager does not carry out plan singlehandedly. It is the manager's job to get other people to do this.

Do Phase

To gain the maximal conceivable savings capability, the investigated action plan should be prioritized and restated into a work plan. And also involve the accountabilities and timeframe for different processes, the work plan must very time contain the essential resources. Highlight the energy aims and short-comings, but it will only be successful if there is enough funds and availability of resources. In addition, Energy Manager must note the achieved targets and analysis the benefits in terms of cost saving. The achieved cost saving and reduction in environment pollution should be demonstrate to motivate the employees.

Check Phase

Check phase aims to supervise the performance (by handling energy audits) in the expressions of energy saving and analyse the set targets. Whenever encounter any loss, it is then important that the root causes are investigated and determined to form correction in order to recognize set objectives. Hence, it is necessary that set objectives are significant to promote the appraisal development.

Act Phase

This phase mostly involves management analysis of audit, internal and external reports referring to the achievement of the energy management program. This information plays a huge part for the organization to spot any loss and act up on it to provide continuous progress.

Environmental, Economic And Social Benefits Of Industrial Energy Efficiency

Environmental

The use of environmentally damaging fossil fuels as a primary energy source has provided to be a difficult issue for the Steam Industry to address. As known that, oil-based fuels have some built-in convenience, density, and portability features that make them particularly easy to transport and especially suitable as a form of energy to power transportation activities. The Steam Industry has taken some important measures to minimize the impact of fossil fuel energy on the environment.

Economic

The principal aim of any energy conservation project is to save money using limited energy or using energy more efficiently. The economic assessment of energy projects can often be complicated since it involves the engineering techniques needed to classify the magnitude of energy savings, together with the economic principles associated in checking in case that capital investment in the project is supported. The utilization of the energy plant is important because if it is pushed beyond the rated flow-sheet capacity, the variable costs may rise rapidly, because of inadequate time for preventive maintenance, resulting in excessive wear or breakdowns and high labour costs for weekend and holiday maintenance work. These are called super-production costs.

Energy Management Challenge

An organisation must have a plan to effectively address any challenges that might be encountered. Sometimes these plans are driven by compliance – to meet statutory requirements and adhere to policies, for instance, health and safety, environmental management, and equal opportunities policies.

Like each of these issues, increasing levels of legislation governing energy consumption, reporting mechanisms and the need for business efficiency mean that facilities managers should acknowledge that energy management also needs similar levels of ownership and responsibility throughout the organization – from board level to the shop floor. In addition, visitors also need to understand how to play their part in reducing unnecessary use of energy. The strategy and plan need active boardroom support or the initiative will stall at the technical level and not result in organizational culture change. The aim should be to ensure that the strategy is embedded in the management practices and becomes the 'new' 'business as usual'.

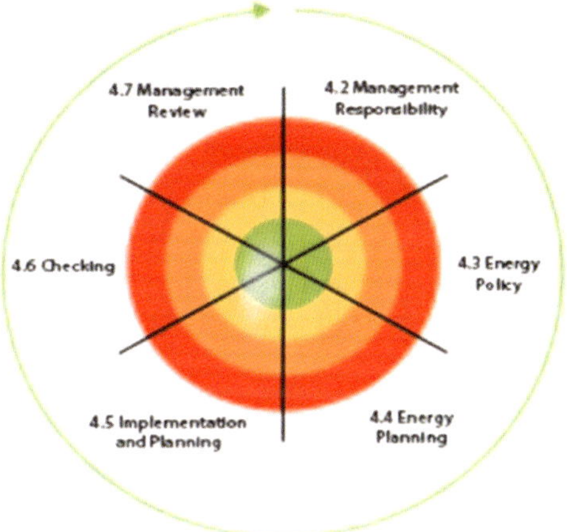

Figure: The components of the continuous process of ISO 50001 that are an integral part of a much wider picture – reproduced from the IET's Guide to Energy Management

System

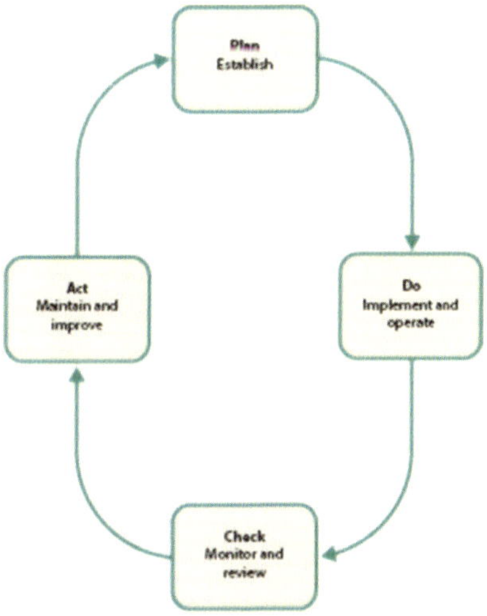

Figure: Plan, Do, Check, Act cycle – reproduced from the IET's Guide to Energy Management

For energy management to succeed there must be a process-driven system to facilitate change supported by technology that closely assesses what, where, why and how the energy is used. Careful monitoring and analysis will identify areas and opportunities for improvement.

The international standard ISO 50001 – *Energy Management* provides a framework for an energy management system using the universal model of 'plan, do, check, act' that is often adapted to manage improvements in the engineering world.

However, facilities managers need to be aware that there is no single solution. A robust energy management system should:

- Have appropriate policies in place and include processes that employ a wide variety of tools, all of which have been adapted for the local needs of the business;
- Reflect strategic direction;
- Include procedures that make the best use of resources;
- Have clear aims and objectives that are quantifiable and are 'smart' in nature; and
- Be flexible enough so that it is adaptable in response to changes in use, changes to the overall size of the estate, or changes to occupancy.

However, it is important to recognize that any changes need careful recording so that data can be compared and realistically analysed.

References

- Energy-management-meaning-steps-and-objectives-6348: businessmanagementideas.com, Retrieved 25 April 2018
- Johannes Kals: Betriebliches Energiemanagement - Eine Einführung, p. 75 – 77, Kohlhammer, Stuttgart 2010, ISBN 978-3-17-021133-9
- Energy-management: energylens.com, Retrieved 10 May 2018
- What-is-energy-management-system-and-its-importance: mhtlighting.com, Retrieved 13 June 2018
- Strategies-for-energy-management: facilityexecutive.com, Retrieved 12 March 2018
- Energy-management-the-foremost-challenge-for-facilities-managers-65: electrical.theiet.org, Retrieved 10 April 2018

6

Energy Conservation

The process of energy consumption achieved through the efficient use and reduction in the consumption of energy is known as energy conservation. All the diverse aspects of energy conservation such as heat rate, rebound effect, fuel saving device, superinsulation, etc. have been carefully analyzed in this chapter.

Energy conservation is one of the words you are hearing more and more. Unfortunately, a lot of the places you will hear it will be in ads marketing products or lifestyle habits that may have nothing to do with actual energy conservation. To learn what real energy conservation techniques are, you have to begin to understand the theory of conservation behind them.

Energy conservation is not about making limited resources last as long as they can, that would mean that you are doing nothing more than prolong a crisis until you finally run out of energy resources all together. Conservation is the process of reducing demand on a limited supply and enabling that supply to begin to rebuild itself. Many times the best way of doing this is to replace the energy used with an alternate.

In the case of fossil fuels, the conservation also can include finding new ways to tap into the Earth's supply so that the commonly used oil fields are not drained completely. This allows for those fields to replenish themselves more. This is not a process that happens overnight, when you are talking about replenishing natural resources you are talking about alleviating excess demand on the supply in 100's of years' time to allow nature to recover.

Importance

Without energy conservation, the world will deplete its natural resources. While some people don't see that as an issue because it will take many decades to happen and they foresee that by the time the natural resource is gone there will be an alternative; the depletion also comes at the cost of creating an enormous destructive waste product that then impacts the rest of life. The goal with energy conservation techniques is reduce demand, protect and replenish supplies, develop and use alternative energy sources, and to clean up the damage from the prior energy processes.

Practical Methods of Energy Conservation

Below are 8 energy conservation techniques that can help you to reduce your overall carbon footprint and save money in the long run.

1. Install CFL Lights: Try replacing incandescent bulbs in your home with CFL bulbs. CFL bulbs cost more upfront but last 12 times longer than regular incandescent bulbs. CFL bulbs will not only save energy but over time you end up saving money.

2. Lower the room temperature: Even a slight decrease in room temperature lets say by only a degree or two, can result in big energy savings. The more the difference between indoor and outdoor temperature, the more energy it consumes to maintain room temperature. A more smarter and comfortable way of doing this is to buy a programmable thermostat.

3. Fix air leaks: Proper insulation will fix air leaks that could be costing you. During winter months, you could be letting out a lot of heat if you do not have a proper insulation. You can fix those leaks yourself or call an energy expert to do it for you.

4. Use maximum daylight: Turn off lights during the day and use daylight as much as possible. This will reduce the burden on the local power grid and save you good amount of money in the long run.

5. Get energy audit done: Getting energy audit done by hiring an energy audit expert for your home is an energy conservation technique that can help you conserve energy and save good amount of money every month. Home energy audit is nothing but a process that helps you to identify areas in your home where it is losing energy and what steps you can take to overcome them. Implement the tips and suggestions given by those energy experts and you might see some drop in your monthly electricity bill.

6. Use energy efficient appliances: When planning to buy some electrical appliances, prefer to buy one with Energy Star rating. Energy efficient appliances with Energy Star rating consume less energy and save you money. They might cost you more in the beginning but it is much more of an investment for you.

7. Drive less, walk more and carpooling: Yet another energy conservation technique is to drive less and walk more. This will not only reduce your carbon footprint but will also keep you healthy as walking is a good exercise. If you go to office by car and many of your colleagues stay nearby, try doing carpooling with them. This will not only bring down your monthly bill you spend on fuel but will also make you socially more active.

8. Switch off appliances when not in use: Electrical appliances like coffee machine, idle printer, desktop computer keep on using electricity even when not in use. Just switch them off if you don't need them immediately.

9. Plant shady landscaping: Shady landscaping outside your home will protect it from intense heat during hot and sunny days and chilly winds during the winter season. This will keep your home cool during summer season and will eventually turn to big savings when you calculate the amount of energy saved at the end of the year.

10. Install energy efficient windows: Some of the older windows installed at our homes aren't energy efficient. Double panel windows and other vinyl frames are much better than single pane windows. Choosing correct blinds can save on your power bills.

Other Energy Conservation Techniques

The other few energy conservation techniques may surprise you. While there are practical methods such as insulation, changing light sources, using alternate fuels and carpooling rather than walking – understand the 6 core techniques beneath them will show you more about what to do in life.

1. Education: Education is probably the most powerful of the energy conservation techniques that can be used. Education is about more than teaching people the importance of conservation, it is about showing the alternative choices that can be used in construction, manufacturing and other processes.

2. Zero Energy Balance: Zero Energy Balance is more than techniques of conserving energy in green construction. It is a process of re-evaluating and retrofitting manufacturing and commercial operations so that they can harvest and store energy, as well as take and replace it onto the grid to relieve brown out stresses.

3. Alternative Power: There are more processes that are starting to use alternative power and fuel sources in many different areas of life. The use of alternative power is one of the most key energy conservation techniques because almost all of the transition models require that the existing processes bc upgraded or replaced to more energy efficient models too.

4. Cap and Trade Agreements: Cap and trade agreements are used as part of the process of regulating and conserving consumption and pollution for manufacturing industries. The companies are "allowed" a certain emission rate which they can bid buy to extend. The extension bid is then used for compensating projects. While this may not seem like it is directly related to energy conservation it is very much at its core.

5. Reduced Demand: There are numerous initiatives that are working to reduce the overall demand on the energy resources of the world. This can range everywhere from education programs to changing the type of required insulation in new construction.

6. Research & Development: Continued funding of research and development projects in the energy conservation field is how we discover the changes that can be made to reduce consumption and discover renewable methods to provide us with the energy that modern life requires. It should be one of the energy conservation techniques that are most valued as it is what holds the promise for leading to a solution to the world's energy crisis.

Barriers to Energy Conservation

(a) Poor Awareness:

Due to lack of understanding of the problem and inadequate information.

(b) Incorrect Attitude:

Arising from the misconception that energy conservation implies deprivation or sacrifice, and the low priority accorded to energy efficiency.

(c) Weak Institutions:

Inadequacy of government as well as private sector organisations in formulating and implementing energy management and conservation policies.

(d) Insufficient Technical Know-how:

Inability to diagnose, design and implement technical solutions to energy efficiency related problems.

(e) Economic and Market Distortions:

Irrational response to conservation measures because of price and other market distortions, or socio-economic factors; and

(f) Capital Shortages:

Inability to finance technically and economically viable energy conservation projects.

Heat Rate

Heat rate is the common measure of system efficiency in a steam power plant. It is defined as "the energy input to a system, typically in Btu/kWh, divided by the electricity generated, in kW." Mathematically:

$$Heat\ Rate(BTU / KWh) = \frac{Input\ Energy\ (BTU / hr)}{Output\ Power\ (KW)}$$

Efficiency is "a ratio of the useful energy output by the system to the energy input to the system." Mathematically:

$$Efficiency = \frac{Useful\ Output\ Energy}{Input\ Energy}$$

As you can see, heat rate is simply the inverse of efficiency. With that in mind, if you increase plant efficiency, which is good, then you would lower heat rate. What this means is that the lower the heat rate, the better your plant is running and, therefore, the more competitive it is.

As you can see by the equations, efficiency has no units, but heat rate does. That is because with heat rate, you typically are measuring the chemical energy input and the electrical energy output, and this will leave you with the units of BTU/kWh.

Plant Heat Rate

Plant heat rate is a measure of the combined performance of the gas turbine cycle, the steam turbine cycle, and any other associated auxiliaries. This may include more than one gas turbine and steam turbine. Heat rate can be further divided to compare units at the same plant, where you might compare Power Block 1 to Power Block 2. You may even use heat rate to compare to other generating units at different sites. Caution must be taken when comparing different types of generating stations, as their chemical energy input may be quite different from yours.

When calculating plant heat rate, the energy input to the system is the chemical energy in the fuel.

Chemical Energy of Fuel = Total Fuel Used (scf/hr) x Higher Heating Value (HHV)(BTU/scf)

The power generated is simply the gross or net generation in kW.

- If gross generation is used, then the resultant heat rate is the gross unit heat rate.

- If net generation is used, then the resultant heat rate is net unit heat rate.

By substituting from the above equations we get:

$$Net\ Unit\ Heat\ Rate\left(\frac{BTU}{KWH}\right)=\frac{Fuel\ Burned(scf\ /\ hr)\times LHV(BTU\ /\ scf)}{Net\ Output\ Power(KW)}$$

Unit analysis is very important. At many sites, fuel flow is measured in lbm/hr, and HHV is given in scf/hr. It is very important to make sure that you match units when doing these problems. At some sites, there is a chromatograph that will measure the heat content of the fuel, and your DCS will convert the fuel flow and HHV to MMBTU. It is very important you make sure you are using the correct units in your equations.

Conversion Factors for Standard Cubic Foot to Pounds Mass

.0458 lbm per scf or 21.826 scf per lbm

Practice Problem:

Calculate plant heat rate for the following conditions (GT CC):

$$FuelFlow=3{,}430{,}000\ scf\ /\ hr$$

$$HHV=1{,}035\ BTU\ /\ scf$$

$$Total\ Plant=480MW(Net)\ and\ 500\ MW$$

$$Output\qquad (Gross)$$

$$Net\ Unit\ HeatRate\left(\frac{BTU}{KWh}\right)=\frac{3{,}430{,}000\ scf\ /\ hr\times 1{,}035\ BTU\ /\ scf}{480{,}000\ KW}$$

$$Net\ Unit\ HeatRate\left(\frac{BTU}{KWh}\right)=7{,}395.94\ BTU\ /\ KWh$$

$$Gross\ Unit\ HeatRate\left(\frac{BTU}{KWh}\right)=\frac{3{,}430{,}000\ scf\ /\ hr\times 1{,}035\ BTU\ /\ scf}{500{,}000\ KW}$$

$$Gross\ Unit\ HeatRate\left(\frac{BTU}{KWh}\right)=7{,}100.10\ BTU\ /\ KWh$$

As you will notice, there is a 4.1% difference between gross heat rate and net heat rate. This shows you the importance of maintaining site loads at a minimum. That will increase your net plant output and, thus, lower your heat rate.

An important point to make here is that the percentage the plant is loaded has a factor on heat rate. The closer the plant is to base load, the better the heat rate. This is due to many factors. The gas turbine is tuned to operate more efficiently at base load. Additionally, the steam turbine will be more efficient at base load. Also, you normally still have the same house loads at minimum load as you do at base load, or they are not proportionally lowered.

It is very important that your site marketing team understands this and understands the consequences financially of dispatching the units in less than full load configuration.

Importance of Heat

Heat rate and thermal performance improvement are integral parts of any serious effort for cost containment in an electric generating station. As the electric power industry expands deregulation and competition, cost containment and the ability to provide energy at the lowest possible cost become important issues. The power producer must provide a lower-cost energy product than the competition and yet still remain profitable in the long term. Fuel-cost reduction and increased reliability and availability through efficiency improvement are key methods of improving profitability.

In many generating stations, fuel expenditure is as high as 90% of the total operations budget. As a result, a 1% reduction in fuel usage achieved through heat rate improvements has a significant, positive impact on profitability. For a 500 MW combined cycle unit, the annual fuel savings could easily exceed $1,500,000 by simply improving heat rate by 200 units.

Heat rate improvement requires the support of personnel at all levels, in addition to support from both station and corporate management. By monitoring and acting on many of the items identified here and decreasing the amount of controllable losses, plant heat rate can be improved to an optimum level and maintained at that optimum level.

Controllable losses, often called operator controllable losses, are defined as "those heat rate losses that can be directly impacted (either positively or negatively) by the actions of the operator." In many cases, the actual "control" is handled by the control system, but often, operator intervention can impact the magnitude of the loss.

Heat Rate Deviation

Most power plants have a target or design heat rate that they are trying to achieve. If the actual heat rate does not match the target, the difference between the actual and target heat rate is the heat rate deviation. Most heat rate deviations are often small in relation to the overall heat rate. For instance, a 70 Btu/kWh deviation is only 1% of a 7,000 Btu/kWh heat rate that is normal for a combined cycle plant.

This does not mean much to most people; however, if we calculate the fuel cost associated with the heat rate deviation, this puts in perspective what this is really costing the plant. Once employees see the dollar figure attached to the losses, it is easier to get all personnel involved in trying to improve heat rate. This is especially important in the corporate budgeting process.

If two projects have been proposed for two different units and each project will improve unit heat rate by 75 Btu/kWh, which project has the best payback? If the units have similar heat rates, similar fuel costs, and are similar in size, the payback will be nearly equal. However, if one unit is 525 MW and the other is 215 MW, then which project has the shortest payback? Without performing the cost calculation, it may be impossible to know. That is one of the primary uses of the heat rate cost calculations. When using heat rate information for making business decisions, it is important to realize that heat rate is only one component in the decision-making process.

Other considerations include system reliability, emissions, maintenance costs, operations costs, and safety concerns. For example, it would not be a good economic decision to install a piece of equipment to improve heat rate if the maintenance cost of the equipment was more than the heat rate savings.

Cost of Heat Rate Deviations

The cost of a heat rate increase can be directly tied to the cost of fuel. Heat rate is expressed in units of Btu/kWh, and fuel is usually purchased in quantities of million Btus. A reduction in heat rate is simply a reduction in the quantity of fuel necessary to produce a fixed number of kWh.

To calculate heat rate deviation cost, it is important to understand that several factors need to be known and understood. Those include gross capacity, auxiliary loads, fuel cost, and the unit's capacity factor. What we are going to calculate is the cost of 1 Btu/kWhr, and then from there, we can determine what savings can be had by an improvement or what something is costing the facility in lost heat rate. We are going to relate this to one year of operation so it can be seen what this cost is per year. This will give you the dramatic look at what savings and losses can be had with minor deviations.

Annual Fuel Cost = Heat Rate Deviation (HRD) x Fuel Cost (FC) x Capacity Factor (CF) x [Gross Capacity (GC) Auxiliary Loads (AL)] x Time (Number of Hours in a Year)

Since we will be calculating for what one unit of heat rate costs, we will substitute 1 BTU/kWhr for our heat rate deviation.

Gas Turbine Combined Cycle Calculation

Other Assumptions:

$$Fuel\ Cost = \$8.50\ /\ MMBTU$$

$$Gross\ Capacity = 525MW$$

$$Auxiliary\ Loads = 15MW$$

$$Capacity\ Factor = 0.25$$

$$Time = 8,760\ hours$$

$$Cost = \frac{1Btu}{KWH} \times \frac{\$8.50}{1,000,000btu} \times 0.25 \times (525\ MW - 15MW) \times \frac{1,000KW}{MW} \times 8,760hr$$

$$Cost = \$9,439.65$$

This is the annual cost for one unit of heat rate deviation. With that said, if you do just a 0.5% increase in deviation on a plant that has a heat rate of 7,400, this will be a 37-unit heat rate improvement, which calculates to an annual fuel savings of $349,267.05. This number usually gets the attention of most of the people at the plant.

You will be able to use this equation when calculating savings for your plant on various projects.

As an example, a project was conducted on a combined cycle facility where several drain valves on the HRSG were leaking during operation. It was determined that this was causing a loss in heat rate due to the loss in steam. The project was to replace the drain valves with better-quality metal-seated ball valves. This project cost was $15,000 per unit. Since the owners were unsure that there would be a cost benefit seen, they would only allow one unit to be done. Once the results on the heat rate improvement were documented, they would decide if it was cost-effective to do the other unit. Once the valves were replaced on the unit, the heat rate on the unit was improved by 11 Btu/kWhr. The cost for one Btu/kWhr was $8,321.

Key Areas in Improving Power Plant Efficiency

A standard coal-fired plant can be broken down into three distinct areas that all contribute to producing heat:

1. Boiler system
2. Turbine
3. Generator

Improving heat rate, and therefore thermal efficiency, means interacting with each of these three systems. Engineers want to increase the efficiency of the boiler, reduce the net heat rate of the turbine system, and boost net generation compared to gross generation.

One of the areas that is most prone to heat rate losses is the boiler system, where the heat from the fuel is converted into steam energy that feeds into the turbine. Within a boiler system, there are a few sources of loss of heat energy, including:

- Sensible heat loss
- Latent heat loss
- Radiation/convection loss
- Unburned combustible loss
- Margin/unknown loss

By examining each source, you can figure out how to improve heat rate by manipulating variables through the application of engineering principles and new equipment. For example, combustion controls that allow for more precise control over excess air levels in the plant's furnace can reduce sensible heat loss (i.e. loss that can be detected via temperature differentials). Lower excess air levels mean the temperature of the exhaust gas is as close as possible to the ambient temperature when the air for combustion enters the plant, therefore resulting in lower rates of loss.

Sensible heat loss can be further reduced by installing air preheaters to raise the ambient temperature of combustion air before it enters the plant.

Tuned boilers – and upgraded boilers in general – can contribute to lower unburned combustible loss rates that can result in a net gain of over 1% in efficiency, which is a significant improvement.

There are ways to further improve efficiency by working on the turbine system. Leakage at the bucket tip and packing level can account for as much as 40% of all the efficiency lost in the turbine. Turbine deposits are another common culprit that can result in significant losses. Other common areas for improvement include nozzles (particularly keeping them from eroding and wearing away) and buckets (also vulnerable to erosion and roughness).

One major change to a turbine system that can result in rapid efficiency improvement is in the turbine blades themselves. Regular maintenance of condensers, heaters, pumps, piping, and cooling towers can reduce efficiency losses, as can upgrading blades with low-pressure variants.

Additionally, replacing the plant's main fan motors with VFD versions (variable frequency drives) can result in heat rate benefit of up to 0.5%. Industrial fan upgrades in general can contribute to better efficiency and constitute one of the most overlooked areas for heat rate improvement.

Rebound Effect

Sustainable resource use necessitates efficient use of energy, raw materials and water. Increased efficiency allows products to be manufactured and services to be performed using fewer resources, and often at a lower cost. This in turn influences purchasing behaviour and product use.

For example, if car prices drop due to efficiency increase, car purchasers are more likely to opt for a larger model the next time they buy a car. A car that's less of a gas guzzler engenders lower per-mile fuel costs, and this usually influences driving behaviour: people tend to use their car more often, drive longer distances, and use mass transit or bicycles less than would otherwise be the case. Hence technically feasible efficiency optimization is often not achieved by virtue of the products in question being used more frequently or more intensively.

Apart from this direct rebound effect (direct changes in product use), other environmentally relevant changes in consumer demand can also occur. For example, the fuel costs saved by car owners might be used instead for air travel (indirect rebound), thus partly negating the energy savings.

How Large is the Rebound

The results of statistical rebound effect projections hinge on the methods used and the effects factored into the calculations. It is particularly difficult to differentiate rebound effects versus effects attributable to economic growth or those occasioned by structural change. Hence such projections vary widely from one study to another. The nature of the products or services being used is also a significant factor. Hence activities such as daily commuting for which time is a limiting factor are subject to a weaker cost rebound effect than are activities that are less time dependent such as flying to Hawaii for vacation. Market saturation also plays an important role, as is evidenced by the fact that rebound effects are far smaller in affluent countries than in developing countries, where pent-up consumer demand tends to be greater.

Types of Effects

The full rebound effect can be distinguished into three different economic reactions to technological changes:

1. Direct rebound effect: An increase in consumption of a good is caused by the lower cost of use. This is caused by the substitution effect.

2. Indirect rebound effect: The lower cost of a service enables increased household consumption of other goods and services. For example, the savings from a more efficient cooling system may be put into another luxury good. This is caused by the income effect.

3. Economy wide effect: The fall in service cost reduces the price of other goods, creates new production possibilities and increases economic growth.

In the example of improved vehicle fuel efficiency, the direct effect would be the increased fuel use from more driving as driving becomes cheaper. The indirect effect would incorporate the increased consumption of other goods enabled by household cost savings from increased fuel efficiency. Since consumption of other goods increases, the embodied fuel used in the production of those goods would increase as well. Finally, the economy-wide effect would include the long-term effect of the increase in vehicle fuel efficiency on production and consumption possibilities throughout the economy, including any effects on economic growth rates.

Direct and Indirect Effects

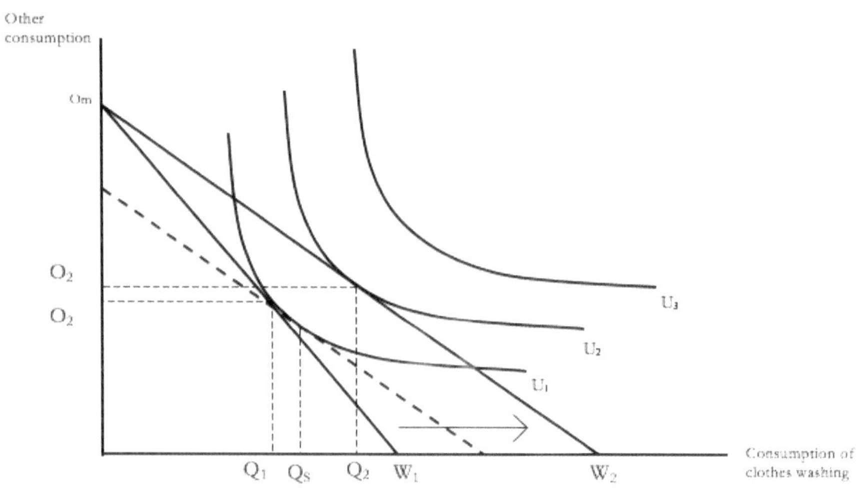

For cost reducing resource efficiency, distinguishing between direct and indirect effects is shown in figure above. The horizontal axis shows units of consumption of the targets good (which could be for example clothes washing, and measured in terms of kilograms of clean clothes) with consumption of all other goods and services on the vertical axis. An economical technology change that enables each unit of washing to be produced with less electricity results in a reduction of the price per unit of washing. This shifts the household budget line rightwards. The result is a substitution effect because of the decreased relative price, but also an income effect due to the increased real income. The substitution effect increases consumption of washing from Q1 to QS, and the income effect from QS to Q2. The total increase in consumption of washing from Q1 to Q2 and the resulting increase in electricity consumption is the direct effect. The indirect effect comprises the increase in other consumption, from O1 to O2. The scale of each of these effects depends on the elasticity of demand for each of the goods, and the embodied resource or externality associated with each good. Indirect effects are difficult to measure empirically. In the manufacturing sector, it has been estimated that there is about a 24% rebound effect due to increases in fuel efficiency.

A parallel effect will happen for cost saving efficient technologies for producers, where output and substitution effects will occur.

The rebound effect can increase the difficulty of projecting the reduction in greenhouse emissions from an improvement in energy efficiency. Estimation of the scale of direct effects on residential electricity, heating and motor fuel consumption has been common motivation for research of rebound effects. Evaluation and econometric methods are the two approaches generally employed in estimating the size of this effect. Evaluation methods rely on quasi-experimental studies and measure the before and after changes to energy consumption from the implementation of energy efficient technology, while econometric methods utilize elasticity estimates to forecast the likely effects from changes in the effective price of energy services.

Research has found that in developed countries, the direct rebound effect is usually small to moderate, ranging from roughly 5% to 40% in residential space heating and cooling. Some of the direct rebound effect can be attributed to consumers who were previously unable to use a service. However, the rebound effect may be more significant in the context of the undeveloped markets in developing economies.

Indirect Effects from Conservation

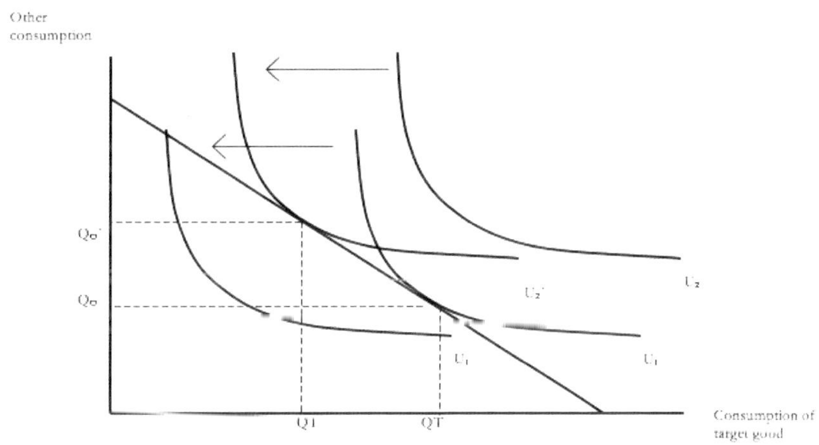

Figure: change in preferences of a household revealing indirect effects from conservation

For conservation measures, indirect effects closely approximate the total economy-wide effect. Conservation measures constitute a change in consumption patterns away from particular targeted goods towards other goods. Figure above shows that a change in preference of a household results in a new consumption pattern that has less of the target good (QT to QT`), and more of all other goods (QO to QO`). The resource consumption or externalities embodied in this other consumption is the indirect effect.

Although a persuasive view has prevailed that indirect effects with respect to energy and greenhouse emissions should be very small due to energy directly comprising only a small component of household expenditure, this view is gradually being eroded. Many recent studies based on life-cycle analysis show the energy consumed indirectly by households is often higher than consumed directly through electricity, gas, and motor fuel, and is a growing proportion. This is evident in the results of recent studies that indicate indirect effects from household conservation can range from

10% to 200% depending on the scenario, with higher indirect rebounds from diet changes aiming to reduce food miles.

Economy Wide Effects

Even if the direct and indirect rebound effects add up to less than 100%, technological improvements that increase efficiency may still result in economy-wide effects that results in increased resource use for the economy as a whole. In particular, this would happen if increased resource efficiency enables an expansion of production in the economy, and an increase in the rate of economic growth. For example, for the case of energy use, more efficient technology is equivalent to a lower price for energy resources. It is well known that changes in energy costs have a large impact on economic growth rates. In the 1970s, sharp increases in petroleum prices led to stagflation (recession and inflation) in the developed countries, whereas in the 1990s lower petroleum prices contributed to higher economic growth. An improvement in energy efficiency has the same effect as lower fuel prices, and leads to faster economic growth. Economists generally believe that especially for the case of energy use, more efficient technologies will lead to increased use, because of this growth effect.

To model the scale of this effect, economists use computational general equilibrium (CGE) models. While CGE methodology is by no means perfect, results indicate that economy-wide rebound effects are likely to be very high, with estimates above 100% being rather common. One simple CGE model has been made available online for use by economists.

Income Level Variation

Research has shown that the direct rebound effects for energy services is lower at high income levels, due to less price sensitivity. Studies have found that own-price elasticity of gas consumption by UK households was two times greater for households in the lowest income decile when compared to the highest decile. Studies have also observed higher rebounds in low-income houses for improvements in heating technology. Evaluation methods have also been used to assess the scale of rebound effects from efficient heating installations in lower income homes in the United Kingdom. This research found that direct effects are close to 100% in many cases. High income households in developed countries are likely to set the temperature at the optimum comfort level, regardless of the cost – therefore any cost reduction does not result in increased heating, for it was already optimal. But low-income households are more price sensitive, and have made thermal sacrifices due to the cost of heating. In this case, a high direct rebound is likely. This analogy can be extended to most household energy consumption.

The size of the rebound effect is likely to be higher in developing countries according to macro-level assessments and case studies. One case study was undertaken in rural India to evaluate the impact of an alternative energy scheme. Households were given solar powered lighting in an attempt to reduce the use of kerosene for lighting to zero except for seasons with insufficient sunshine. The scheme was also designed to encourage a future willingness to pay for efficient lighting. The results were surprising, with high direct rebounds between 50 and 80%, and total direct and indirect rebound above 100%. Because the new lighting source was essentially zero cost, operating hours for lighting went up from an average of 2 to 6 per day, with new lighting consisting of a combination of both the no-cost solar lamps and also kerosene lamps. Also, more cooking was undertaken which enabled an increased trade of food with neighboring villages.

Rebounds with Respect to Time

The individual opportunity of cost is an often overlooked cause of the rebound effect. Just as improved workplace tools result in an increased expectation of productivity, so does the increased availability of time result in an increase in demand for a service. Research articles often examine increasingly convenient and more rapid modes of transportation to determine the rebound effect in energy demand. Because time cost forms a major part of the total cost of commuter transport, rapid modes will reduce real costs, but will also encourage longer commuting distances which will in turn increase energy consumption. While important, it is almost impossible to estimate empirically the scale of such effects due to the subjective nature of the value of time. Time saved can either be used towards additional work or leisure which may have differing degrees of rebound effect. Labor time saved at work due to the increased labour productivity is likely to be spent on further labor time at higher productive rates. For leisure time saving, this may simply encourage people to diversify their leisure interests to fill their generally fixed period of leisure time.

Suggested Solutions

In order to ensure that efficiency enhancing technological improvements actually reduce fuel use, the ecological economists Mathis Wackernagel and William Rees have suggested that any cost savings from efficiency gains be "taxed away or otherwise removed from further economic circulation. Preferably they should be captured for reinvestment in natural capital rehabilitation." This can be achieved through, for example, the imposition of a green tax, a cap and trade program, higher fuel taxes or the proposed "restore" approach where part of the savings is directed back to the resource. Policies can also directly address projected yearly consumption of energy rather than device efficiency, especially for systems where the use can be accurately projected, such as street lighting.

Fuel Saving Device

Fuel saving device is an accessory device and/or a consumable which allows to decrease the consumption of fuel of an existing heat engine for an identical usage.

The accessory character of the fuel-saving device and its simplicity make it an interesting solution to improve what exists.

It holds no connection to the good practices of the user regarding:

- The driving of the engine

- Its maintenance and the maintenance of its peripheral devices

- The choice of a vehicle adapted to its use

It should be reasonably easy to install not to enter into the category of a significant alteration, as described in administrative terminology, which modifies the conception of the engine. When made standard equipment, it becomes an innovation and should be described as part of the performance of the manufacturer.

Processes used to Decrease the Consumption

Antistatic Treatment of The Fuel

The aim is to remove the static charges accumulated by the fuel which, by landing up on the droplets, attract the droplets to the wall of the cylinder. The mix becomes less homogeneous and does not burn as well. Antistatic agents (for example STADIS ® 425) are used in fuels and solvents to disperse the static electricity. We have all seen unsightly antistatic black rubber bands rubbing the rears of cars. Another technique is to trap the static charges by diverting them with a magnetic field, just before the engine.

Balancing Static Charges Between Air and Fuel

In order to improve an antistatic treatment of the fuel, you should add charges to the air. It can be done in a simple way, linking the body to a sharp stuff pricked through the air intake (Point effect needle).

Water Enhancement

Simply adding water allows to improve the quality of the combustion. Water is an indispensable catalyst in most chemical reactions which, without dampness, see their kinetics vary in a considerable way. In the presence of heavy hydrocarbon and in some conditions, water produces a cracking into lighter elements of which combustion becomes easier than that of the initial types found in the engine cylinder. Water also has a direct thermodynamic impact on combustion due to the enthalpies of phase change (latent heat).

Water can either be mixed to intake air or to the fuel (aquazole).

Addition of catalysed damp air to the air intake

The technique improves the performance of water enhancement. A catalytic reactor heated by the exhaust gases improves the physicochemical reactions in the presence of water and air, which would not have happened in the case of a direct intake, especially reforming when carbon is available, and static charges balancing (steam easily gets charged and gives its charges through a sharp diffusor by point effect).

Injection of Water-alcohol

In diesel engines, injection of a water-alcohol mixture (sometimes called aquaol) really consists in producing an alcohol bi-fuelling. But the presence of water also acts on combustion and improves performance. Compressing a tank which contains a water-alcohol mixture with the pressure of the turbo and injecting it with a very fine nozzle after the turbo is sufficient. Without any specific tuning, the system is only triggered when the engine is under load.

Addition of Brown Gas to the Air Intake

Brown gas, also called HHO, is a monatomic mixture of hydrogen and oxygen in a gaseous state. It is produced by the pulsed electrolysis of an aqueous electrolyte (water and potassium hydroxide for instance). Its molar volume is twice that of the dihydrogen and dioxygen mixture obtained by means of a classic electrolysis. Mixed with the air intake, it quickly recombines to water at the beginning of the combustion of the gas volume, producing its known effects. The current intake from the alternator should not be too high, as it would then produce the reverse effect :reduced mileage.

Body Aerodynamics

Adding a deflector or taking the carrier bars off the roof are well known solutions. Other exists, such as the implementation of vanes at the rear of the vehicle (a rather ugly solution) to optimise the path of the boundary layer. The surface state of the body of the car also plays a role, which is difficult to assess or to master, however.

Aerodynamics of the Intake Circuit

The path of the air between the filter and the engine should be made easy to minimise frictional losses. Replacing the piping may bring an improvement, therefore. If the air intake is located at the front of the car, dynamic pressure of the relative wind also plays a role and improves the filling rate. In the case of "petrol" engines, having the air intake rotate after its carburisation improves carburisation by a brewing process, but this should not slow down the air beyond a certain threshold. Propeller-like devices called AVEC made with cut metal sheets are actually charges diffusors using the point effect.

Intercooler

An intercooler is an air-to-air or air-to-water exchanger which allows to cool the air intake (whether carburated or not) and to improve the filling rate. Output is increased and consumption is reduced. Added to an old engine, this works as a fuel-saving device.

Turbo Charger

The turbo compresses the air intake to increase the filling rate by using the speed of exhaust gases as a source of energy. As is the case with the intercooler, this can be considered a fuel-saving device when added to an old engine, what is not easy to achieve.

Improvement of Carburettors

Many devices have been invented and commercialised to improve carburettors:

- Richness correction device

- Turbulators to improve the mixture

- Water injectors

We believe that special carburettors (of which there is a long list) are not true fuel-saving devices as everything needs to be changed.

Recycling of the Fuel Vapours

On modern petrol vehicles, the "canister" device fills this function. Collector's cars would highly benefit from such an equipment, even without a true canister.

Electronic Microchips

Changing the microchip of a calculator is similar to changing the programme which manages the fuel injection (cartography). On some engines of a single series, manufacturers offer several powers simply by implementing different programmes. It is easy to understand that an optimisation is possible and that, in some cases, consumption can be reduced.

Bio-fuels

Using bio-fuels does not allow to reduce fuel consumption per se, but it does limit the consumption of fossil fuels. Vegetal oil or bio-ethanol (SP95 E10 or E85) can be mentioned, but they have the counter effect of monopolising agriculture for other objectives than food. Gasifier synthesized water gas is more eco-friendly, but calls for an irreproachable organisation. Recycled oils, mixed to diesel oil in small proportions, are a very good recycling chain. Bio-gas produced by methanisation of waste would be ideal, but it happens to be extremely corrosive for engines.

Usage of Gas or of Bio-gas in Diesel Engines

The process consists in letting the injection pump in idle position while accelerating the engine and mixing gas (propane, butane, bio-gas and so forth) with the air intake. The economical result is better in certain conditions. The process holds its place on this page, therefore, as a faraway cousin.

Improvement of Ignition

In the case of "petrol" positive-ignition engines, this is a crucial factor. The second spark plug of plane engines works as a fuel-saving device if one considers that it secures the spark. Electronic ignition on an old vehicle has a spectacular effect and should be the first investment one makes on a collector's car which is not equipped. Such ignition devices now take advantage of the flexibility of electronic, which allows multiple configurations.

Usage of spark plugs with built-in capacitors also offers an appreciated benefit by a considerable improvement of the quality of the spark. This technology is a standard feature of the french light helicopter Guimbal Cabri G2, it is mentioned in the datasheet of the aircraft, powered by a Textron Lycoming O360-J2A.

Intake Vortex Devices

Another device with staunch critics, intake vortex devices are based on the principle the combustion of all fossil fuels has one common element: oxygen. The ratio between a fossil fuel and oxygen is called the stoichiometric ratio.

An intake cortex device is a mechanism designed to change the stoichiometric ratio of a fuel to a cleaner burn by increasing the sum of oxygen in the mix. The industry standard for gasoline is 14.7 parts oxygen to 1 part fuel. For diesel, the stoichiometric ratio is 14.5:1.

There was a time when intake vortex devices did have an effect on fuel economy. However, running a lean fuel mixture has its disadvantages as well. For one, an engine runs hot on a lean mixture. That can lead to problems like gasket failure, metal stress and fatigue, and a general shortening of the life of an engine.

A lean mixture also means low octane. The combustion of low octane fuels can lead to engine pings and knocking if not bigger issues. Even if an intake vortex device were to work as advertised, it would mean taking a chance of damaging an engine by running too lean a fuel.

But, intake vortex devices no longer work as advertised because the fuel to air mixture is controlled by a computerized sensor in the electronic fuel injection (EFI) of modern vehicles, trucks, and heavy equipment. If the air to fuel mixture is lean, an onboard computer changes the mixture to a richer ratio.

And, there is a question as to what agitating air prior to it mixing with fuel actually does. The distance from the intake to the piston cylinder is choked valves and corners and constricting hoses that work against any notion creating a "mini-tornado' in the combustion chamber.

Fuel Ionizers

Debunking the fuel savings claims made by the manufacturers of intake vortex devices and hydrogen generators require a scientific explanation if they are — in fact — a hoax. It is more difficult to debunk fuel ionizers because some manufacturers provide no scientific explanation as to why an ionizer increases fuel economy.

The following is an example of a fuel ionizer advertisement:

"The atoms of the crystal materials within the fuel ionizer have been altered by a unique and special process. The fuel ionizer is in high resonance or in tune with higher energies that create a field of "positive energy" that can pass through any type of material. When attached to the fuel line of an engine, the ionizer begins immediately to ionize the fuel, causing the fuel to become more volatile. The result is a fuel that burns more completely, resulting in better fuel mileage, better performance, and helps to reduce emissions."

Whether fuel ionizers are a legitimate fuel saving device or not, the marketing is not particularly convincing.

Fuel-line Magnets

There seems to be a greater amount of convincing and experimentally backed support for magnetic fuel ionization methods.

Fuel and Oil Additives

It would be extremely convenient if fuel and oil additives and treatments improved the fuel efficiency of a vehicle. However, they most certainly don't. For one — irrespective of the marketing label given to an additive or treatment, they are not catalysts. Unlike noble metals and magnets and natures other true catalysts, fuel additives are simply solvent that burns up with the fuel with which they are mixed.

Another reason fuel additives and treatments cannot improve the fuel efficiency of a vehicle is because they cannot break the laws of physics.

To improve the fuel economy of a fuel — which is what some dishonest manufacturers of additives and treatments claim they do — additives and treatments must do one of two things.

An additive or treatment must either boost the energy density of a fuel as a supplement or catalyze a change in the fuel that increases its energy density. Additives and treatments do neither.

The purpose of fuel additives and treatments is to accomplish one of four things or a combination. Fuel additives can clean the internal combustion components of an engine as a detergent. They can increase the cetane/octane of a fuel. Fuel can lubricate the internal components of an engine. Or, an additive can increase fuel flow rates in cold temperature conditions.

What additives and treatments do not do is increase fuel efficiency, aka, fuel economy or "gas" mileage.

Fuel Catalysts

Made of noble and precious metals, fuel catalysts are akin to catalytic converters in that they use true catalysts to both increase fuel efficiency and reduce emissions. Fuel catalysts are in-line, pre-combustion mechanisms cylinders that affect the physical makeup of fuels.

All fossil fuels are heterogeneous mixtures made of clumps of fuel clusters. Fuel clusters are the reason catalytic converters are necessary. The molecules inside a fuel cluster do not have exposure to air. As a result of the lack of exposed surface area, some of the fuel molecules inside a cluster do not combust and, instead, escape out the exhaust unburned.

Fuel molecules cluster as the result of polarization. The inherent charge in molecules causes them to attract one another and cluster together. The noble metals in a fuel catalyst neutralize the charge of fuel molecules which leads to those in a cluster drifting apart. Once separated, fuel molecules can oxygenize, then combust.

What separates fuel catalysts from other fuel saving devices — regardless of whether they work or not is science and cost. The cost of producing fuel catalysts is exponentially higher than that of other fuel saving devices. With the amount of money necessary to manufacture fuel catalysts, most fuel catalyst manufacturers invest large sums of money in research, design, development, and testing.

While it is difficult to determine which fuel saving devices are legitimate and which are merely cow magnets, a little research begins to make it clear which are backed by scientific research and data and which can be made, "from an old peanut-butter jar and some leftover copper pipe or roof flashing."

Air Bleed Devices

Devices have been marketed which bleed a small amount of air into the fuel line before the carburetor. These may improve fuel economy because the engine runs slightly lean as a consequence. However, running leaner than the manufacturer intended can cause overheating, piston damage, loss of maximum power and poor emissions (e.g., higher NOx due to higher combustion temperatures, or, if misfiring occurs, higher HC).

Electronic Devices

Some electronic devices are marketed as fuel savers. The Fuel Doctor FD-47, for example, plugs into the vehicle's cigarette lighter and displays several LEDs. It is claimed to increase vehicle fuel economy by up to 25% through "power conditioning of the vehicle's electrical systems", but *Consumer Reports* detected no difference in economy or power in tests on ten separate vehicles, finding that the device did nothing but light up. *Car and Driver* magazine found that the device contains nothing buts "a simple circuit board for the LED lights", and disassembly and circuit analysis reached the same conclusion. The maker disputed claims that the device has no effect, and proposed changes to the *Consumer Reports* testing procedure, which when implemented made no difference to the results.

Another device described as 'electronic' is the 'Electronic Engine Ionizer Fuel Saver'. Testing of this device resulted in a loss of power and an engine compartment fire.

There are also genuinely useful 'emissions-control defeat devices' that operate by allowing a vehicle's engine to operate outside government-imposed tailpipe emissions parameters. These government standards force factory engines to operate outside their most efficient range of operation. Either engine control units are reprogrammed to operate more efficiently, or sensors that influence the ECU's operation are modified or 'simulated' to cause it to operate in a more efficient manner. Oxygen sensor simulators allow fuel-economy reducing catalytic converters to be removed. Such devices are often sold for "off-road use only".

Thermodynamic Efficiency

The reason why most devices are not capable of producing the claimed improvements is based in thermodynamics. This formula expresses the theoretical efficiency of a petrol engine:

$$\eta = 1 - \frac{1}{r_v^{\gamma-1}}$$

where η is efficiency, r_v is the compression ratio, and γ is the ratio of the specific heats of the cylinder gases.

Assuming an ideal engine with no friction, perfect insulation, perfect combustion, a compression ratio of 10:1, and a 'γ' of 1.27 (for gasoline-air combustion), the theoretical efficiency of the engine would be 46%.

For example, if an automobile typically gets 20 miles<32.19 km> per gallon with a 20% efficient engine that has a 10:1 compression ratio, a carburetor claiming 100MPG would have to increase the efficiency by a factor of 5, to 100%. This is clearly beyond what is theoretically or practically possible. A similar claim of 300MPG for any vehicle would require an engine (in this particular case) that is 300% efficient, which violates the first law of thermodynamics.

Extremely efficient vehicle designs capable of achieving 100MPG+ (such as the VW 1l) do not have substantially greater engine efficiency, but instead focus on better aerodynamics, reduced vehicle weight, and using energy that would otherwise be dissipated as heat during braking

Good Practices of the User Regarding:

- the driving of the engine
- its maintenance and the maintenance of its peripheral devices
- the choice of a vehicle adapted to its use

Driving of the Engine

Some simple rules allow to decrease consumption:

Let us remember that fuel delivers a power (force multiplied by speed) through the engine, power meant to:

- Accelerate

- Go up-hill

- Fight against the friction of wheels and air, more or less proportional to the square of the velocity.

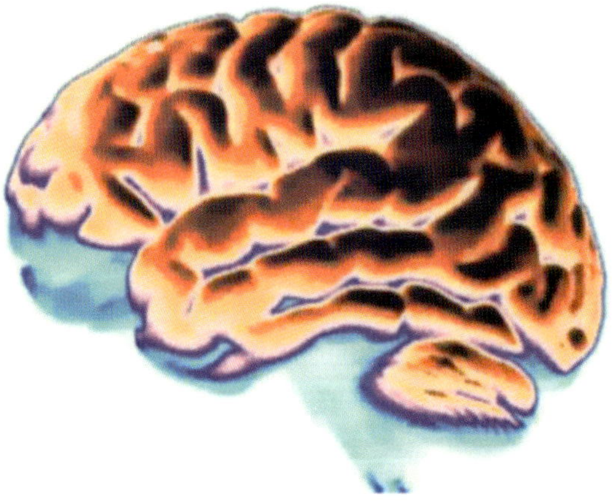

Upper Fuel Saver

Lower Fuel Saver

It can henceforth be deducted that accelerations should be reduced (while taking the range of better performance into consideration) and that one should:

- accelerate slowly when entering a highway

- try not to use the brakes, such a behavioural trick calling for some forward thinking on speed changes, by looking far ahead

- avoid entering into races

- keep a constant speed in general.

It can also be deducted that hilly roads should be avoided (difficult to achieve in a mountain environment)

- As one always goes down at one point, use the slope, without giving up safety.

- The lighter the load, the lesser the consumption ; think about unloading those cement bags from the trunk!

It can lastly be deducted that friction is an important factor, and that one should:

- drive more slowly

- have well inflated tyres

- ahave a clean car

- avoid carrier bars, open windows, antennas, and so forth.

Nota Bene

With a modern car, it is better to remain engaged when going downhill or decelerating as the calculator is going to cut the fuel. When declutching there will be some consumption to keep the idle position. On older vehicles ,the cut does not occur and it is then better to declutch, while remaining very alert as the engine brake is not active any more. Always be ready to clutch and brake.

Maintenance

A well maintained device has a reduced consumption and pollution:

- Clean air filter
- Regular oil-changes with a quality oil, if possible with a change of filter each time.
- Well inflated tyres
- Controlled parallelism
- Quality fuel
- Regular cleaning of the fuel circuit by a specialist (flexible and injector)
- During Winter, let the engine warm up for a few minutes, if possible when the vehicle is stopped.

Superinsulation

Retrofits

It is possible, and increasingly desirable, to retrofit superinsulation to an existing houses or buildings. The easiest way is often to add layers of continuous rigid exterior insulation, and sometimes by building new exterior walls that allow more space for insulation. A vapor barrier can be installed on the outside of the original framing but may not be needed. An improved continuous air barrier is almost always worth adding, as older homes tend to be leaky, and such an air barrier can be important for energy savings and durability. Care should be exercised when adding a vapor barrier as it can reduce drying of incidental moisture, or even cause summer (in climates with humid summers) interstitial condensation and consequent mold and mildew. This may cause health problems for the occupants and damage the existing structure. Many builders in northern Canada use a simple 1/3 to 2/3 approach, placing the vapor barrier no further out than 1/3 of the R-value of the insulated portion of the wall. This method is generally valid for interior walls that have little or no vapor resistance (e.g. they use fibrous insulation) and controls air leakage condensation as well as vapor diffusion condensation. This approach will ensure that condensation does not occur on or to the inside of the vapor barrier during cold weather. The 1/3:2/3 rule will ensure that the vapor barrier temperature will not fall below the dew point temperature of the interior air, and will minimize the possibility of cold-weather condensation problems. For example, with an internal room temperature of 20°C (68°F), the vapor barrier will then only reach 7.3 °C (45 °F) when the outside temperatures is at –18°C (-1°F). Indoor air dewpoint temperatures are more likely to be in the order of around 0 °C (32 °F) when it is that cold outdoors, much lower than the predicted vapor barrier temperature, and hence the 1/3:2/3 rules is quite conservative. For climates that do not often experience -18°C, the 1/3:2/3 rule should be amended to 40:60% or 50:50. As the interior air dewpoint temperature is an important basis for such rules, buildings with high interior humidities during cold weather (e.g., museums, swimming pools, humdified or poorly ventilated airtight homes) may require different rules, as can buildings with drier interior environments (such as highly ventilated buildings, warehouses). The 2009 International Residential Code (IRC) embodies more sophisticated rules to guide the choice of insulation on the exterior of new homes, which can be applied when retrofitting older homes.

A vapor permeable building wrap on the outside of the original wall helps keep the wind out, yet allows the wall assembly to dry to the exterior. Asphalt felt and other products such as permeable polymer based products are available for this purpose, and usually double as the Water Resistant Barrier / drainage plane as well.

Interior retrofits are possible where the owner wants to preserve the old exterior siding, or where setback requirements don't leave space for an exterior retrofit. Sealing the air barrier is more difficult and the thermal insulation continuity compromised (because of the many partition, floor, and service penetrations), the original wall assembly is rendered colder in cold weather (and hence more prone to condensation and slower to dry), occupants are exposed to major disruptions, and the house is left with less interior space. Another approach is to use the 1/3 to 2/3 method mentioned above — that is, to install a vapor retarder on the inside of the existing wall (if there isn't one there already) and add insulation and support structure to the inside. This way, utilities (power, telephone, cable, and plumbing) can be added in this new wall space without penetrating the air barrier. Polyethylene vapor barriers are risky except in very cold climates, because they limit the wall's ability to dry to the interior. This approach also limits the amount of interior insulation that can be added to a rather small amount (e.g., only R6 can be added to a 2x4 R12 wall).

Costs and Benefits

In new construction, the cost of the extra insulation and wall framing may be offset by not requiring a dedicated central heating system. In homes with numerous rooms, more than one floor, air conditioning or large sized, a central furnace is often justified or required to ensure sufficiently uniform temperatures. Small furnaces are not very expensive and some ductwork to every room is almost always required to provide ventilation air in any case. When peak demand and annual energy use is low, sophisticated and expensive central heating systems are not often required. Hence, even electric resistance heaters may be used. Electric heaters are typically only used on the coldest winter nights when overall demand for electricity is low. Other forms of backup heater are widely used, such as wood pellets, wood stoves, natural gas boilers or even furnaces. The cost of a superinsulation retrofit should be balanced against the future cost of heating fuel (which can be expected to fluctuate from year to year due to supply problems, natural disasters or geopolitical events), the desire to reduce pollution from heating a building, or the desire to provide exceptional thermal comfort.

During a power failure, a superinsulated house stays warm longer as heat loss is much less than normal, but the thermal storage capacity of the structural materials and contents is the same. Adverse weather may hamper efforts to restore power, leading to outages lasting a week or more. When deprived of their continuous supply of electricity (either for heat directly, or to operate gas-fired furnaces), conventional houses cool rapidly during, and may be at greater risk of costly damage due to freezing water pipes. Residents who use supplemental heating methods without proper care during such episodes, or at any other time, may subject themselves to risk of fire or carbon monoxide poisoning.

Energy Recycling

Energy recycling is the recovery of energy that would normally be wasted in industrial processes by flaring, exhausting to the atmosphere or operating low efficiency equipment, and converting it into electricity or thermal energy (steam). Combined heat and power (CHP) is a form of energy recycling, where power generation facilities are designed to produce energy and also supply heat rather than waste it. Energy recycling and CHP can be implemented at industrial sites, manufacturing facilities and large institutions such as hospitals and universities.

In a typical electric power generation plant, input fuel is used to create electricity while excess thermal energy (in the form of steam) is wasted in the process. Many facilities, both commercial and industrial, require a substantial amount of thermal energy for heating, cooling and other low temperature processes. By locating an electric generation plant on-site at a facility that has a significant demand for thermal energy, steam that is typically wasted can be used by the host. Each CHP plant reduces their host's energy costs and reduces emissions as compared to buying power off of the grid and producing steam with boilers.

Recycling waste energy streams is accomplished by Primary Energy in two ways:

- converting industrial waste energy streams into heat and power with on-site generation plants; and
- building combined heat and power ("CHP") facilities near thermal energy users to enable recycling of normally wasted thermal energy from the production of electric power to displace host boiler fuel.

Forms of Energy Recycling

Waste heat recovery is a process that captures excess heat that would normally be discharged at manufacturing facilities and converts it into electricity and steam, or returns energy to the manufacturing process in the form of heated air, water, glycol, or oil. A "waste heat recovery boiler" contains a series of water-filled tubes placed throughout the area where heat is released. When high-temperature heat meets the boiler, steam is produced, which in turn powers a turbine that creates electricity. This process is similar to that of other fired boilers, but in this case, waste heat replaces a traditional flame. No fossil fuels are used in this process. Metals, glass, pulp and paper, silicon and other production plants are typical locations where waste heat recovery can be effective.

Waste heat recovery from air conditioning is also used as an alternative to wasting heat to the atmosphere from chiller plants. Heat recovered in summer from chiller plants is stored in Thermalbanks in the ground and recycled back to the same building in winter via a heat pump to provide heating without burning fossil fuels. This elegant approach saves energy - and carbon - in both seasons by recycling summer heat for winter use.

Combined heat and power (CHP), also called cogeneration, is, according to the U.S. Environmental Protection Agency, "an efficient, clean, and reliable approach to generating electricity and heat energy from a single fuel source. By installing a CHP system designed to meet the thermal and electrical base loads of a facility, CHP can greatly increase the facility's operational efficiency and decrease energy costs. At the same time, CHP reduces the emission of greenhouse gases, which contribute to global climate change." When electricity is produced on-site with a CHP plant, excess heat is recycled to produce both processed heat and additional power.

Enabling technologies: Heat pumps and thermal energy storage are classes of technologies that can enable the recycling of energy that would otherwise be inaccessible due to a temperature that is too low for utilization or a time lag between when the energy is available and when it is needed. While enhancing the temperature of available renewable thermal energy, heat pumps have the additional property of leveraging electrical power (or in some cases mechanical or thermal power) by using it to extract additional energy from a low quality source (such as seawater, lake water, the ground, the air, or waste heat from a process).

Thermal storage technologies allow heat or cold to be stored for periods of time ranging from hours or overnight to interseasonal, and can involve storage of sensible energy (i.e. by changing the temperature of a medium) or latent energy (i.e. through phase changes of a medium, such between water and slush or ice). Short-term thermal storages can be used for peak-shaving in district heating or electrical distribution systems. Kinds of renewable or alternative energy sources that can be enabled include natural energy (e.g. collected via solar-thermal collectors, or dry cooling towers used to collect winter's cold), waste energy (e.g. from HVAC equipment, industrial processes or power plants), or surplus energy (e.g. as seasonally from hyropower projects or intermittently from wind farms). The Drake Landing Solar Community (Alberta, Canada) is illustrative. borehole thermal energy storage allows the community to get 97% of its year-round heat from solar collectors on the garage roofs, which most of the heat collected in summer. Types of storages for sensible energy include insulated tanks, borehole clusters in substrates ranging from gravel to bedrock, deep aquifers, or shallow lined pits that are insulated on top. Some types of storage are capable of storing heat or cold between opposing seasons (particularly if very large), and some storage applications require inclusion of a heat pump. Latent heat is typically stored in ice tanks or what are called phase-change materials (PCMs).

Electric Turbo Compounding (ETC)

Electric Turbo Compounding (ETC) is a technology solution to the challenge of improving energy efficiency for the stationary power generation industry.

Fossil fuel based power generation is predicted to continue for decades, especially in developing economies. This is against the global need to reduce carbon emissions, of which, a high percentage is produced by the power sector worldwide.

ETC works by making gas and diesel-powered gensets (Electric Generators) work more effectively and cleaner, by recovering waste energy from the exhaust to improve power density and fuel efficiency.

Advantages of using ETC

- Helps developing economies with unreliable or insufficient power infrastructure.

- Gives independent power providers (IPPs), power rental companies and generator OEMs (original equipment manufacturers) a competitive advantage and potential increased market share.
- Improves overall efficiency of the genset, including fuel input costs and helping end-users reduce amount of fuel burned.
- Typically 4-7% less fuel consumption for both diesel and gas gensets.
- Fewer carbon emissions.
- Increased power density.
- Capability to increase power output and capacity, with improved fuel efficiency.
- ETC system integration offers a step change in efficiency without increasing service or maintenance requirements.
- The cost of generating power through waste heat recovery is substantially less than burning more fuel, even with low diesel prices.

Disadvantages of using ETC

- Upfront costs incur an additional expense for businesses.
- The need to update existing turbo machinery and recertification of the unit adds additional costs and can be time consuming.
- There will be additional weight to add an ETC to a current unit.
- Process still uses fossil fuels, thus still has a carbon footprint in a renewable age.
- They are bespoke to each generator so the design, build and implementation can be a lengthy process.
- There are challenges with high speed turbo generators such as high stress in the rotors, heat generation of the electrical machine and rotor dynamics of the turbo generator system.

Current System

Both waste heat recovery and CHP constitute "decentralized" energy production, which is in contrast to traditional "centralized" power generated at large power plants run by regional utilities. The "centralized" system has an average efficiency of 34 percent, requiring about three units of fuel to produce one unit of power. By capturing both heat and power, CHP and waste heat recovery projects have higher efficiencies.

A 2007 Department of Energy study found the potential for 135,000 megawatts of CHP in the U.S., and a Lawrence Berkley National Laboratory study identified about 64,000 megawatts that could be obtained from industrial waste energy, not counting CHP. These studies suggest about 200,000 megawatts—or 20% - of total power capacity that could come from energy recycling in the U.S. Widespread use of energy recycling could therefore reduce global warming emissions by an estimated 20 percent. Indeed, as of 2005, about 42 percent of U.S. greenhouse gas pollution came from the production of electricity and 27 percent from the production of heat.

Advocates contend that recycled energy costs less and has lower emissions than most other energy options in current use.

Currently Recycling Energy Int. Corp. takes advantage of recycling energy in heat recovery ventilation and latent heat pump and CHCP.

Radiator Reflector

Radiator reflectors work on a very simple principle. You radiator radiates heat both into the room, and into the wall behind it. That heat can simply be lost to the outside, especially if you have thin single skin walls. Reflectors are installed behind the radiator and help prevent that heat from being lost by reflecting the heat back into the room.

Reasons for not using foil Radiator Reflector

There are a couple of reasons to use an approved material. This approved material does not crinkle and crease like standard kitchen foil – it stays shiny and flat, in the ideal position to reflect heat. Foil can also oxidise over time as it is exposed to the environment, and this reduces its reflective properties, whereas radiator reflectors are treated with a chemical to prevent oxidisation. Putting kitchen foil behind the radiator is a false economy, as its effectiveness will be severely limited by these effects over time.

Homemade foil radiator reflectors crinkle easily and oxidise, losing their reflective qualities

Radiators that Work with Radiator Reflectors

The walls where Radiator reflector will make the most difference is in exposed areas of the property where the radiator is situated on an exterior wall. Internal walls are not heat loss walls, and therefore you won't save any heat by installing reflectors there. Party walls between properties normally don't need reflectors either, unless your next door neighbours don't heat the house, or there is no one living there.

The type of radiator itself is usually unimportant, as reflectors will work with almost any radiator, as long as there is a space to fit the reflector in behind it.

References

- Shurcliff, William A., Superinsulated Houses and Air-To-Air Heat Exchangers, Brick House Pub Co, 1988, ISBN 0-931790-73-5
- Energy-conservation-techniques: conserve-energy-future.com, Retrieved 12 March 2018
- Saunders, Harry D. (2008). "Fuel conserving (and using) production functions". Energy Economics. 30 (5): 2184–2235. doi:10.1016/j.eneco.2007.11.006
- Barriers-to-energy-conservation-activities-29309: yourarticlelibrary.com, Retrieved 10 June 2018
- Jevons, William Stanley (1866). The Coal Question (2nd ed.). London: Macmillan and Co. Chapter VII. Retrieved 2008-07-21.
- Increase-heat-rate-coal-power-plants: processbarron.com, Retrieved 25 April 2018
- Saunders, Harry (1992). "The Khazzom-Brookes Postulate and Neoclassical Growth". Energy Journal. 13: 131–148. doi:10.5547/issn0195-6574-ej-vol13-no4-7
- Fuel-saving-device-works-doesnt-anyones-guess: rentar.com, Retrieved 09 July 2018
- Wackernagel, Mathis; Rees, William (1997). "Perceptual and structural barriers to investing in natural capital: economics from an ecological footprint perspective". Ecological Economics. 20 (3): 3–24. doi:10.1016/S0921-8009(96)00077-8.

7

Renewable Energy

The energy produced from renewable resources which is easily replenished is called renewable energy. Some sources of renewable energy are wind, sunlight, rain, tides, geothermal heat, etc. The topics included in this chapter on wind power, biomass, hydropower, solar energy, etc. discuss the major renewable energies and their generation.

Renewable energy is a term for any useable energy that is harnessed from natural resources that are either essentially inexhaustible (such as sunlight, or thermal energy generated and stored in the Earth) or naturally replenished in a timely manner on a human timescale (such as energy derived from wood). In contrast, non-renewable energy refers to energy derived from resources that are limited in amount and cannot be replenished in a a timely manner or feasibly and are essentially irreplaceable once extracted (such as energy derived from fossil fuels). The term renewable energy has not only been applied to the energy and forms of energy harnessed but also to the sources and technologies associated with that usable energy. Examples of renewable energy technologies include methods to harness water power, sunlight, thermal energy, or biomass for the generation of electricity.

The term renewable energy is often used interchangeable with alternative energy and green energy. While by most definitions there is substantial overlap between energy forms, sources, and technologies that fit into these three categories, the three terms also have been delineated differently. The term alternative energy generally references any nontraditional energy form, source, or technology differing from the current popular forms, sources, or technologies. Before natural gas gained popularity, this energy source could be classified under the category of alternative energy; however, it does not fit under renewable energy. Green energy references that subset of renewable energy that involves the least environmental harm.

The movement toward renewable energy reflects not just an awareness of the need to prepare for a future when fossil fuels are no longer in plentiful supply, but also the importance of balancing economic development with protection of the environment for future generations.

The wind, sun, and biomass are three renewable energy sources.

Burbo Bank Offshore Wind Farm, at the entrance to the River Mersey in North West England.

There are a multitude of definitions used for renewable energy. The various conceptions allow debate regarding what is included in this category, whether energy from peat, large hydropower plants, or biomass are counted as renewable energies, and whether one includes nuclear power.

A commonly cited definition is that provided by the International Energy Agency (IEA 2008)

"Renewable energy is derived from natural processes that are replenished constantly. In its various forms, it derives directly or indirectly from the sun, or from heat generated deep within the earth. Included in the definition is energy generated from solar, wind, biomass, geothermal, hydropower and ocean resources, and biofuels and hydrogen derived from renewable resource.

This definition is used by the United Nations Environment Programme (UNEP 2011). The IEA website references a similar but differently worded version: "Energy derived from natural processes (e.g. sunlight and wind) that are replenished at a faster rate than they are consumed. Solar, wind, geothermal, hydro, and some forms of biomass are common sources of renewable energy" (IEA 2014).

A definition used by the European Union utilizes this conception: "Renewable energy sources are defined as renewable non-fossil energy sources: wind, solar, geothermal, wave, tidal, hydropower, biomass, landfill gas, sewage treatment plant gas and biogases".

The German Renewable Energy Act (Erneuerbare-Energien-Gesetz or EEG) delineates what fits into the renewable energy category as "hydropower including wave power, tidal power, salt gradient and flow energy, wind energy, solar radiation, geothermal energy, energy from biomass including biogas, landfill gas and sewage treatment plant gas, as well as the biodegradable fraction of municipal and industrial waste." Not included within this definition is biomass from non-biodegradable plants nor large hydropower plants.

A panoramic view of the Whitelee Wind Farm with Lochgoin Reservoir in the foreground.

Michael Hoexter offers this analysis: "Non-renewable energy sources are energy stores with zero or a minute rate of replenishment relative to its depletion by human beings" and "renewable energy sources are types of natural energy flux useful for human ends regularly occurring on or near Earth's surface and, additionally, useful natural energy stores that are replenished by natural flux within the time frame of conceivable human use." Hoexter further "all known renewable energy sources originate in, or are close derivatives of, electromagnetic radiation of our Sun, the Earth's and Moon's gravitational fields, and heat radiating from Earth's interior.

Whether a particular resources serves as a source of renewable energy is often a point of contention. Some countries, such as Sweden and Finland, label peat as a renewable source of energy, while the World Energy Council (WEC) suggests to not consider peat a renewable source of energy. Peat has a notably slow and low renewable rate. Some US and UK politicians have termed nuclear power a renewable energy, and physicist Bernard Cohen has stated that uranium is "practically inexhaustible" and can be considered a renewable course of energy, with naturally-replenished uranium extracted from seawater able to supply energy as long as the sun's expected lifespan. Other have advocated against inclusion of nuclear energy and some has even questioned geothermal since it may lead to partial depletion at some locales.

The term renewable energy forms, technologies, and sources—is distinct from the term renewable natural resource in that this later term generally also includes resources from which useable energy is not obtained, such as fish stocks or wildlife (and non-renewable resources include minerals, such as gold, silver, diamonds, and copper). An energy source, according to IEA/Eurostat definition is "the kinetic (e.g. hydro, wind), thermal (eg. nuclear, geothermal), or combustible fuel used as the input to generate electricity or heat". Energy inside a system is transformed from one energy form to another (thermal to electricity), and an energy technology is that which transforms, transports, or stores the energy form.

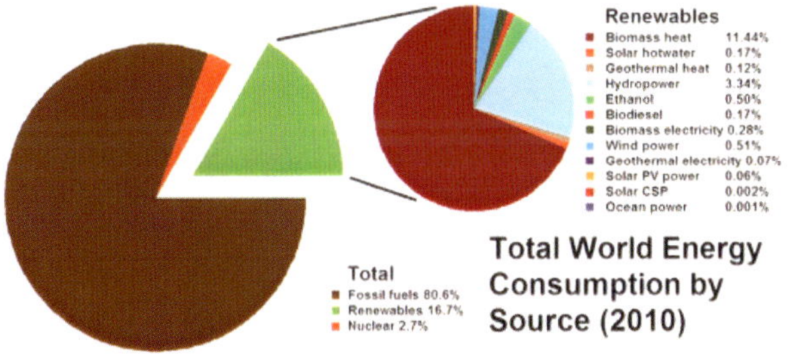

In 2010 renewable energy accounted for 17% of total energy consumption.
Biomass heat accounted for 11%, and hydropower 3%

In 2010, renewable energy provided an estimated 16.7% of global final energy consumption, according to the Renewable Energy Policy Network for the 21st Century (REN21). About 8.5% of this came from traditional biomass and about 8.2% came from modern renewable energy (hydropower, solar, wind, geothermal, biofuels, modern biomass). Hydropower accounted for about 3.3% of global final energy consumption in 2010 (REN21 2012).

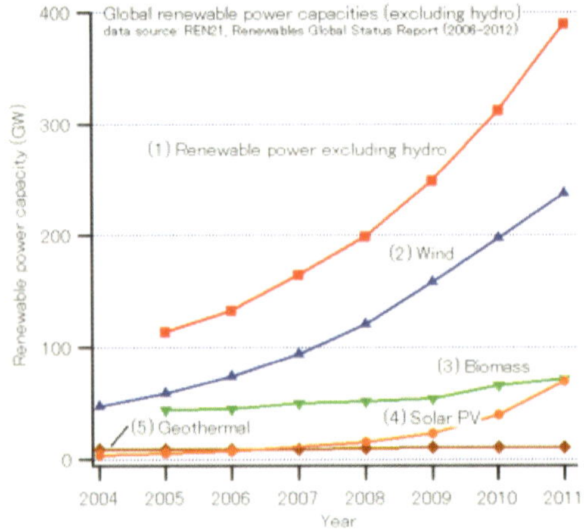

Global renewable power capacity excluding hydro.

The International Energy Agency (IEA) reported that in 2010 the use of renewable energy (including traditional biomass) accounted for about 13% of global primary energy demand (IEA 2012) and in 2009 renewables accounted for 19.5% of global electricity generation (IEA 2014). Bloomberg reported that renewable sources, including large hydro, had a 12.6% share of total primary energy production in 2010.

At the national level, by 2013 at least 30 nations around the world had renewable energy contributing more than 20% of energy supply. National renewable energy markets are projected to continue to grow strongly in the coming decade and beyond (REN21 2013). Wind power, for example, is growing at the rate of 30% annually, with a worldwide installed capacity of 282,482 megawatts (MW) at the end of 2012.

The 150 MW Andasol Solar Power Station is a commercial parabolic trough solar thermal power plant, located in Spain. The Andasol plant uses tanks of molten salt to store solar energy so that it can continue generating electricity even when the sun isn't shining.

In the United States, renewable energy sources were used in 2012 to provide abut 12% of total U.S.

utility-scale electricity generation, mostly from hydroelectic power (56%), followed by wind (28%), biomass wood (8%), biomass waste (4%), geothermal (3%), and solar (1%) (EIA 2013). The top nations in terms of total electricity generation from renewable energy, in order, are China, United States, Brazil, and Canada, while the order for production of electricity from non-hydroelectric renewable sources are the United States, China, and Germany (EIA 2013).

Renewable energy resources and significant opportunities for energy efficiency exist over wide geographical areas, in contrast to other energy sources, which are concentrated in a limited number of countries. Renewable energy replaces conventional fuels in four distinct areas: electricity generation, hot water/space heating, motor fuels, and rural (off-grid) energy services.

Renewable Technologies

Geothermal Energy

In general terms, geothermal energy is thermal energy (the energy that determines the temperature of matter) generated and stored in the Earth. The geothermal energy of the Earth's crust originates from the original formation of the planet and from radioactive decay of minerals, resulting in continual production of geothermal energy below the earth's surface. The geothermal gradient, which is the difference in temperature between the core of the planet and its surface, drives a continuous conduction of thermal energy in the form of heat from the core to the surface.

In terms of renewable energy, geothermal energy is the energy that is harnessed from the Earth's internal heat and used for practical purposes, such as heating buildings or generating electricity. It also refers to the technology for converting geothermal energy into useable energy. The term geothermal power is used synonymously as the conversion of the Earth's internal heat into a useful form of energy, or more specifically as the generation of electricity from this thermal energy (geothermal electricity).

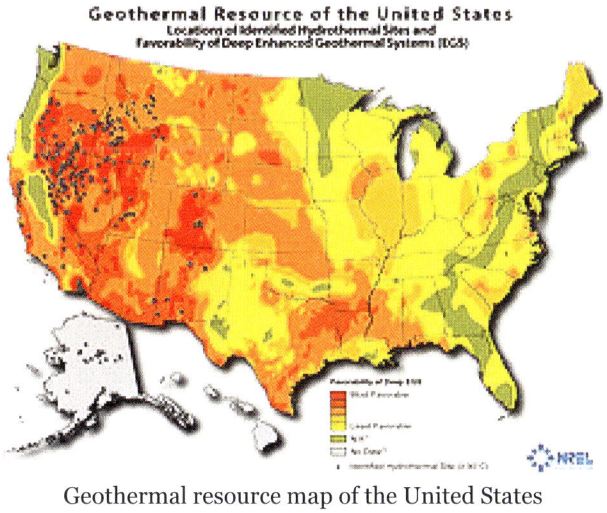

Geothermal resource map of the United States

The four basic means for capturing geothermal energy for practical use are geothermal power plants (dry steam, flash steam, binary cycle), geothermal heat pumps, direct use, and enhanced geothermal systems.

Geothermal provides a huge, reliable, renewable resource, unaffected by changing weather conditions. It reduces reliance on fossil fuels and their inherent price unpredictability, and when managed with sensitivity to the site capacity, it is sustainable. Furthermore, technological advances have dramatically expanded the range and size of viable resources.

Steam rising from the Nesjavellir Geothermal Power Station in Iceland

However, geothermal also faces challenges in the need for significant capital investment, and a significant amount of time in terms of building geothermal plants. There are limitations in terms of placement of geothermal plants in regions with accessible deposits of high temperature ground water, and construction of power plants can adversely affect land stability. Geothermal power plants also can lead to undesirable emissions, with power plant emitting low levels of carbon dioxide, nitric oxide, sulfur, methane, and hot water from geothermal sources may hold in solution trace amount of toxic elements, such as mercury, boron, and arsenic.

From hot springs, geothermal energy has been used for bathing since Paleolithic times and for space heating since ancient Roman times, but it is now better known for electricity generation.

Wind Power

Generally speaking, wind energy is the form of energy created by wind. In terms of alternative energy, wind energy refers to the energy that is harnessed from wind for practical purposes. The term wind power is used synonymously as the conversion of wind energy into a useful form of energy, or more specifically as the generation of electricity from the wind. Among ways in which wind energy can be harnessed are wind turbines to make electrical power, windmills for mechanical power, windpumps for water pumping or drainage, or sails to propel ships.

Large wind farms consist of hundreds of individual wind turbines that are connected to the electric power transmission network. For new constructions, onshore wind is an relative inexpensive source of electricity, while small onshore wind farms provide electricity to isolated locations. Modern utility-scale wind turbines range from around 600 kW to 5 MW of rated power, although turbines with rated output of 1.5–3 MW have become the most common for commercial use; the power available from the wind is a function of the cube of the wind speed, so as wind speed increases, power output increases dramatically up to the maximum output for the particular turbine. Areas where winds are stronger and more constant, such as offshore and high altitude sites, are preferred locations for wind farms. As offshore wind speeds average ~90% greater than that of land, offshore resources can contribute substantially more energy than land stationed turbines.

The Shepherds Flat Wind Farm is a 845 megawatt (MW) wind farm in the U.S. state of Oregon.

Wind power offers a number of benefits as an alternative to fossil fuels. It is plentiful, renewable, widely distributed, clean, produces no greenhouse gas emissions during operation, and uses little land. The effects on the environment are generally less problematic than those from other power sources. Costs are relatively low and once the infrastructure is paid for it is virtually free.

The main disadvantage of wind power is the fact that wind is unpredictable, inconsistent, and unsteady, as well as the concern that the full costs of harnessing wind power are not cheap and thus rely on government subsidies to be set up and be competitive. There also are aesthetic concerns, with wind farms being considered by some to be an eyesore, whether restricting the normally picturesque view offshore or in rural areas. Furthermore, there are complaints of noise from turbines, and some communities have been required to shut off their turbines during certain times because of the noise. Older type wind farms have turbines that spin at high speeds and can thus kill wild birds and bats, although this design has changed so newer wind farms largely avoid such a problem.

As of 2011, Denmark is generating more than a quarter of its electricity from wind and 83 countries around the world are using wind power to supply the electricity grid. In 2010 wind energy production was over 2.5% of total worldwide electricity usage, and growing rapidly at more than 25% per annum. The long-term technical potential of wind energy is believed to be five times total current global energy production, or 40 times current electricity demand, assuming all practical barriers are overcome. This would require wind turbines to be installed over large areas, particularly in areas of higher wind resources, such as offshore.

Hydropower

The energy of falling water and running water can be utilized to provide water power or hydropower—the form of renewable energy derived from the gravitational force of falling or flowing water harnessed for useful purposes. Since water is about 800 times denser than air, even a slow flowing stream of water, or moderate sea swell, can yield considerable amounts of energy. Since ancient times, hydropower has been used for irrigation and the operation of various mechanical devices, such as watermills, sawmills, textile mills, dock cranes, domestic lifts, and power houses.

Since the early 20th century, the term hydropower has been used almost exclusively in conjunction with the modern development of hydroelectric power, which allowed use of distant energy sources. Hydroelectricity is the term referring to electricity generated by hydropower; the production of electrical power through the use of the gravitational force of falling or flowing water.

Seen from space; Three Gorges Dam (left), Gezhouba Dam (right).

There are many forms of water power:

- Hydroelectric energy from large-scale hydroelectric dams. The largest example of a large-scale hydroelectric dam is the Three Gorges Dam in China and a smaller example is the Akosombo Dam in Ghana.
- Micro hydro systems are hydroelectric power installations that typically produce up to 100 kW of power. They are often used in water rich areas as a remote-area power supply (RAPS).
- Run-of-the-river hydroelectricity systems derive kinetic energy from rivers and oceans without the creation of a large reservoir.
- Trompe is another method used to transmit energy. A trompe is a water-powered gas compressor, commonly used before the advent of the electric-powered compressor, which is somewhat like an airlift pump working in reverse. A trompe produces compressed air from falling water. Compressed air could then be piped to power other machinery at a distance from the waterfall.

Hydroelectricity today is the most widely used form of renewable energy (unless all biomass categories, such as wood and biofuels, are lumped together), accounting for 16 percent of global electricity generation—3,427 terawatt-hours of electricity production in 2010. Hydropower is produced in at least 150 countries, with five countries (China, Brazil, United States, Canada, and Russia) accounting for about 52 percent of the world's installed hydropower capacity in 2010. The Asia-Pacific region generated 32 percent of global hydropower in 2010, with China being the largest hydropower producer, producing 721 terawatt-hours in 2010 and having the highest installed hydropower capacity, with 213 gigawatts (GW) at the end of 2010. The Three Gorges Dam, spanning China's Yangtze River, is the world's largest hydroelectric power station in terms of installed capacity, followed in second place by the Itaipu Dam in Brazil/Paraguay. The Three Gorges Dam is operated jointly with the much smaller Gezhouba Dam.

The cost of hydroelectricity is relatively low, making it a competitive source of renewable electricity. The average cost of electricity from a hydro plant larger than 10 megawatts is 3 to 5 U.S. cents per kilowatt-hour. Hydro is also a flexible source of electricity since plants can be ramped up and down very quickly to adapt to changing energy demands. Once a hydroelectric complex is constructed, the project produces no direct waste, and has a considerably lower output level of the greenhouse gas carbon dioxide (CO_2) than fossil fuel powered energy plants. However, damming interrupts the flow of rivers and can harm local ecosystems, and building large dams and reservoirs often involves displacing people and wildlife. Given such concerns, in some nations building new dams on major rivers to capture hydroelectric energy meets a lot of resistance and further expansion of hydropower in the United States is unlikely. On the other hand, China's Three Gorges Dam became fully functional in just 2012.

Solar Energy

Broadly speaking, solar energy is energy from the Sun. In terms of renewable energy, solar energy refers to the energy that is harnessed from solar radiation, using the radiant light and heat from the Sun for practical purposes. This energy is harnessed using a range of ever-evolving technologies such as solar photovoltaics, solar heating, solar thermal electricity, solar architecture, and artificial photosynthesis (IEA 2011; RSC 2014). Solar energy can be harnessed at different levels around the world, mostly depending on distance from the equator.

The term solar power either is used synonymously with solar energy or is used more specifically to refer to the conversion of sunlight into electricity. The conversion into electricity is done either directly using photovoltaics (PV), or indirectly using concentrated solar power (CSP). Concentrated solar power systems use lenses or mirrors and tracking systems to focus a large area of sunlight into a small beam. Commercial concentrated solar power plants were first developed in the 1980s. Photovoltaics convert light into electric current using the photoelectric effect. Photovoltaics are an important and relatively inexpensive source of electrical energy where grid power is inconvenient, unreasonably expensive to connect, or simply unavailable. However, as the cost of solar electricity is falling, solar power is also increasingly being used even in grid-connected situations as a way to feed low-carbon energy into the grid.

Solar technologies are broadly characterized as either passive solar or active solar depending on the way they capture, convert and distribute solar energy. Active solar techniques include the use of photovoltaic panels and solar thermal collectors to harness the energy. Passive solar techniques include orienting a building to the Sun, selecting materials with favorable thermal mass or light dispersing properties, and designing spaces that naturally circulate air. Other broad classifications are direct solar, which generally refers to technologies or effects that involve a single-step conversion of sunlight that results in a usable form of energy, and indirect solar, which generally refers to technologies or effects that involve multiple-step transformations of sunlight.

Benefits of solar energy system include the huge potential in terms of energy hitting the earth, the low environmental impact, and the lack of producing carbon dioxide and air pollutants. Limitations preventing the large scale implementation of solar powered energy generation is the inefficiency of current solar technology and the cost. In addition, the amount of sunlight varies depending on weather conditions, location, time of day, and time of year, and the need for a large surface to collect the energy, since it does not deliver concentrated energy at any one place (IEA 2014).

Part of the 354 MW SEGS solar complex in northern San Bernardino County, California.

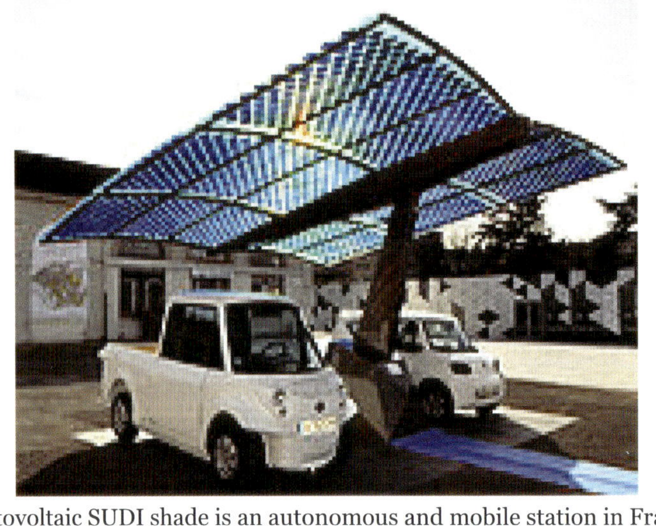

Photovoltaic SUDI shade is an autonomous and mobile station in France that provides energy for electric vehicles using solar energy.

Biomass

Biomass refers to biological material derived from living or recently living organisms. It most often refers to plants or plant-derived materials, which are specifically called lignocellulosic biomass. As an energy source, biomass can either be used directly via combustion to produce heat, or indirectly after converting it to various forms of biofuel. Conversion of biomass to biofuel can be achieved by different methods, which are broadly classified into: thermal, chemical, and biochemical methods. This biomass conversion can result in fuel in solid, liquid, or gas form.

Biofuel, wood, and waste are the three main categories of the use of biomass as an energy source.

Wood remains the largest biomass energy source today; examples include forest residues (such as dead trees, branches, and tree stumps), yard clippings, wood chips, and even municipal solid waste. Industrial biomass can be grown from numerous types of plants, including miscanthus, switchgrass, hemp, corn, poplar, willow, sorghum, sugarcane, bamboo, and a variety of tree species, ranging from eucalyptus to oil palm (palm oil).

Plant energy is produced by crops specifically grown for use as fuel that offer high biomass output per hectare with low input energy. Some examples of these plants are wheat and straw. The grain can be used for liquid transportation fuels, while the straw can be burned to produce heat or electricity. Plant biomass can also be degraded from cellulose to glucose through a series of chemical treatments, and the resulting sugar can then be used as a first generation biofuel.

Biomass can be converted to other usable forms of energy like methane gas or transportation fuels like ethanol and biodiesel. Rotting garbage, and agricultural and human waste, all release methane gas—also called “landfill gas” or “biogas.” Crops, such as corn and sugar cane, can be fermented to produce the transportation fuel, ethanol. Biodiesel, another transportation fuel, can be produced from left-over food products like vegetable oils and animal fats. Also, biomass to liquids (BTLs) and cellulosic ethanol are still under research.

There is a great deal of research involving algal, or algae-derived, biomass due to the fact that it’s a non-food resource and can be produced at rates 5 to 10 times those of other types of land-based agriculture, such as corn and soy. Once harvested, it can be fermented to produce biofuels such as ethanol, butanol, and methane, as well as biodiesel and hydrogen.

The biomass used for electricity generation varies by region. Forest by-products, such as wood residues, are common in the United States. Agricultural waste is common in Mauritius (sugar cane residue) and Southeast Asia (rice husks). Animal husbandry residues, such as poultry litter, are common in the UK.

The generation of renewable energy from biomass on the scale needed to replace fossil energy would present serious environmental challenges. For example, biomass energy generation would have to increase 7-fold to supply current primary energy demand, and up to 40-fold by 2100 given economic and energy growth projections. Humans already appropriate 30 to 40% of all photosynthetically fixed carbon worldwide, indicating that expansion of additional biomass harvesting is likely to stress ecosystems, in some cases precipitating collapse and extinction of animal species that have been deprived of vital food sources. The total amount of energy capture by vegetation in the United States each year is around 58 quads (61.5 EJ), about half of which is already harvested as agricultural crops and forest products. The remaining biomass is needed to maintain ecosystem functions and diversity. Since annual energy use in the United States is ca. 100 quads, biomass energy could supply only a very small fraction. To supply the current worldwide energy demand solely with biomass would require more than 10% of the Earth’s land surface, which is comparable to the area use for all of world agriculture (i.e., ca. 1500 million hectares), indicating that further expansion of biomass energy generation will be difficult without precipitating an ethical conflict, given current world hunger statistics, over growing plants for biofuel versus food. Of course, developing productive means to produce energy from the parts of vegetation not utilized in food production, such as discarded, inedible cellulose components, would help address some of these difficulties.

Sugarcane plantation in Brazil (State of São Paulo), cane remains used to production of biomass energy

A cogeneration plant in Metz, France. The station uses waste wood biomass as energy source, and provides electricity and heat for 30,000 dwellings

Biofuel

Biofuel is one of the three main categories of the use of biomass as an energy source (the others being wood and waste).

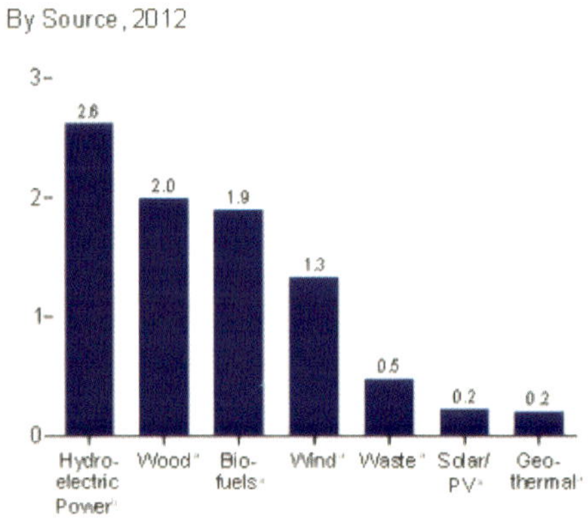

US Renewable Energy Consumption (quadrillion BTU)

A biofuel is a solid, liquid, or gaseous fuel made from biomass. In other words, these fuels contain energy from geologically recent carbon fixation of living or recently living organisms. As noted by Wilkie (2013), "Any combustible fuel derived from recent (non-fossil) living matter (biomass) may be considered a biofuel, including ethanol derived from plant products, biodiesel from plant or animal oils, as well as, biogas from biomass.

Liquid biofuels include bioalcohols, such as bioethanol, and oils, such as biodiesel. Gaseous biofuels include biogas, landfill gas, and synthetic gas.

Brazil has bioethanol made from sugarcane available throughout the country. Shown a typical Petrobras gas station at São Paulo with dual fuel service, marked A for alcohol (ethanol) and G for gasoline.

Bioalcohols: Biologically produced alcohols, most commonly ethanol, and less commonly propanol and butanol, are produced by the action of microorganisms and enzymes through the fermentation of sugars or starches (easiest), or cellulose (which is more difficult). Alcohol fuels are produced by fermentation of sugars derived from wheat, corn, sugar beets, sugar cane, molasses, and any sugar or starch from which alcoholic beverages can be made (such as potato and fruit waste, etc.). These alcohols are also produced by chemical means. When obtained from biological materials and/or biological processes, they are known as bioalcohols (e.g. "bioethanol"). There is no chemical difference between biologically produced and chemically produced alcohols.

Ethanol fuel, or bioethanol, is the most common biofuel worldwide, particularly in Brazil, but also in the United States and elsewhere. Ethanol can be used as a fuel for vehicles in its pure form, but it is usually used as a gasoline additive to increase octane and improve vehicle emissions. With advanced technology being developed, cellulosic biomass, such as trees and grasses, are also used as feedstocks for ethanol production. Cellulosic biofuels, such as cellulosic ethanol, began to be produced in commercial-scale plants in 2013. Most methanol (the simplest alcohol) is produced from natural gas, a nonrenewable fossil fuel, and modern methanol also is produced in a catalytic industrial process directly from carbon monoxide, carbon dioxide, and hydrogen. However, methanol also can be produced from biomass (as biomethanol) using very similar chemical processes. Butanol can be produced from biomass (as "biobutanol")as well as fossil fuels (as "petrobutanol"); but biobutanol and petrobutanol have the same chemical properties.

Biodiesel: Biodiesel is made from vegetable oils, animal fats or recycled greases. Biodiesel can be used as a fuel for vehicles in its pure form, but it is usually used as a diesel additive to reduce levels of particulates, carbon monoxide, and hydrocarbons from diesel-powered vehicles. Biodiesel is produced from oils or fats using transesterification and is the most common biofuel in Europe.

Biogas: Biogas, which is also known as biomethane, landfill gas, swamp gas, and digester gas, is a collection of gases (largely methane and carbon dioxide) produced by the anaerobic degradation of biomass (non-fossil organic matter) by various bacteria. The primary component of biogas is methane gas, which comprises 50-90% by volume of biogas. More than half of the gas used in Sweden to power the natural gas vehicles is biogas.

Biofuels have increased in popularity because of rising oil prices and the need for energy security. In 2010, worldwide biofuel production reached 105 billion liters (28 billion gallons US), up 17% from 2009. Global ethanol fuel production reached 86 billion liters (23 billion gallons US) in 2010, with the United States and Brazil as the world's top producers, accounting together for 90% of global production. The world's largest biodiesel producer is the European Union, accounting for 53% of all biodiesel production in 2010 (Worldwatch 2011). Biofuels provided 2.7% of the world's transport fuel in 2010.

Emerging Technologies

Other renewable energy technologies are still under development, including ocean energy, as well as the above-mentioned cellulosic ethanol and enhanced geothermal systems. These technologies are not yet widely demonstrated or have limited commercialization. Many are on the horizon and may have potential comparable to other renewable energy technologies, but still depend on attracting sufficient attention and research, development, and demonstration (R&D) funding.

There are numerous organizations within the academic, federal, and commercial sectors conducting large scale advanced research in the field of renewable energy. This research spans several areas of focus across the renewable energy spectrum. Most of the research is targeted at improving efficiency and increasing overall energy yields. Two of the most prominent of these labs are Sandia National Laboratories and the National Renewable Energy Laboratory (NREL), both of which are funded by the United States Department of Energy and supported by various corporate partners.

Marine Energy

Marine energy (also sometimes referred to as ocean energy) refers to the energy carried by ocean waves, tides, salinity, and ocean temperature differences. The movement of water in the world's oceans creates a vast store of kinetic energy, or energy in motion. This energy can be harnessed to generate electricity to power homes, transport, and industries.

The term marine energy encompasses both wave energy and tidal energy. Offshore wind power is not a form of marine energy, as wind power is derived directly from the wind, even if the wind turbines are placed over water.

Wave energy is the transport of energy by ocean surface waves. Waves are generated by wind passing over the surface of the sea. As a renewable energy, wave energy is the capture of the energy of waves for a useful purpose, such as electricity generation, water desalination, or the pumping of water (into reservoirs). The term wave power is used synonymously with wave energy, or references the generation of electricity from the energy of waves. Wave power may be considered a form of hydropower whereby the definition of hydropower is expanded to encompass any type of energy gained from the movement of water. Wave power is distinct from the diurnal flux of tidal power and the steady gyre of ocean currents. Wave-power generation is not currently a widely employed commercial technology, although there have been attempts to use it since at least 1890. In 2008, the first experimental wave farm was opened in Portugal, at the Aguçadoura Wave Park.

Rance, second largest tidal power station at 240 MW

Tidal energy is the form of energy created by movement of tides. In terms of renewable energy, tidal energy refers to the energy that is harnessed from the tides for practical purposes. The term tidal power is used synonymously as the conversion of tidal energy into a useful form of energy, or more specifically as the generation of electricity from the tides. Tidal power is the only technology that draws on energy inherent in the orbital characteristics of the Earth–Moon system, and to a lesser extent in the Earth–Sun system. As with wave power, tidal power may be considered a form of hydropower whereby the definition of hydropower is expanded to encompass any type of energy gained from the movement of water. Although not yet widely used, tidal power has potential for future electricity generation. Tidal power has traditionally suffered from relatively high cost and limited availability of sites with sufficiently high tidal ranges or flow velocities, thus constricting its total availability. However, many recent technological developments and improvements, both in design (e.g. dynamic tidal power, tidal lagoons), and turbine technology (e.g. new axial turbines, cross flow turbines), indicate that the total availability of tidal power may be much higher than previously assumed, and that economic and environmental costs may be brought down to competitive levels. The world's first large-scale tidal power plant (the Rance Tidal Power Station) became operational in 1966.

The oceans have a tremendous amount of energy and are close to many if not most concentrated populations. Ocean energy has the potential of providing a substantial amount of new renewable energy around the world.

Lightning

Since the late 1980s, there have been several attempts to investigate the possibility of harvesting energy from lightning. While a single bolt of lightning carries a relatively large amount of energy (approximately 5 billion joules (IOP 2014), this energy is concentrated in a small location and is passed during an extremely short period of time (milliseconds); therefore, extremely high electrical power is involved. It has been proposed that the energy contained in lightning be used to generate hydrogen from water, or to harness the energy from rapid heating of water due to lightning, or to use inductors spaced far enough away so that a safe fraction of the energy might be captured.

A technology capable of harvesting lightning energy would need to be able to rapidly capture the high power involved in a lightning bolt. Several schemes have been proposed, but the ever-changing energy involved in each lightning bolt have rendered lightning power harvesting from ground based rods impractical. Additionally, lightning is sporadic, and therefore energy would have to be collected and stored; it is difficult to convert high-voltage electrical power to the lower-voltage power that can be stored. Another major challenge when attempting to harvest energy from lightning is the impossibility of predicting when and where thunderstorms will occur. Even during a storm, it is very difficult to tell where exactly lightning will strike.

Carbon-neutral and Negative Fuels

Carbon-neutral fuels are synthetic fuels produced by hydrogenating waste carbon dioxide recycled from power plant flue-gas emissions, recovered from automotive exhaust gas, or derived from carbonic acid in seawater. Among such fuels is included methane, gasoline, diesel fuel, jet fuel or ammonia. Production of such fuels is considered to be carbon-neutral because it does not result in a net increase in atmospheric greenhouse gases. To the extent that synthetic fuels displace fossil fuels, or if they are produced from waste carbon or seawater carbonic acid, and their combustion is subject to carbon capture at the flue or exhaust pipe, they result in negative carbon dioxide emission and net carbon dioxide removal from the atmosphere, and thus constitute a form of greenhouse gas remediation.

Such renewable fuels alleviate the costs and dependency issues of imported fossil fuels without requiring either electrification of the vehicle fleet or conversion to hydrogen or other fuels, enabling continued compatible and affordable vehicles. Carbon-neutral fuels offer relatively low cost energy storage, alleviating the problems of wind and solar intermittency, and they enable distribution of wind, water, and solar power through existing natural gas pipelines. Nighttime wind power is considered the most economical form of electrical power with which to synthesize fuel, because the load curve for electricity peaks sharply during the warmest hours of the day, but wind tends to blow slightly more at night than during the day, so, the price of nighttime wind power is often much less expensive than any alternative. Germany has built a 250 kilowatt synthetic methane plant which they are scaling up to 10 megawatts.

The George Olah carbon dioxide recycling plant in Grindavík, Iceland has been producing 2 million liters of methanol transportation fuel per year from flue exhaust of the Svartsengi Power Station since 2011.

Artificial Photosynthesis

Artificial photosynthesis uses techniques include nanotechnology to store solar electromagnetic energy in chemical bonds by splitting water to produce hydrogen and then using carbon dioxide to make methanol. Researchers in this field are striving to design molecular mimics of photosynthesis that utilize a wider region of the solar spectrum and employ catalytic systems made from abundant, inexpensive materials that are robust, readily repaired, non-toxic, stable in a variety of environmental conditions, and perform more efficiently, allowing a greater proportion of photon energy to end up in the storage compounds, i.e., carbohydrates (rather than building and sustaining living cells.

Wind Power

Wind power is the kinetic energy of wind, harnessed and redirected to perform a task mechanically or to generate electrical power.

Sailing and similar activities are among the simplest, oldest and most direct applications of wind power. Mechanical uses of wind power go back to ancient Greece. The first wind wheels, for example, relied upon gearing and shafts to power machinery. Windmills drive a millstone to grind grains, among other purposes. Wind has also long been used to power mechanical pumps to draw water for use and to drain land for agricultural purposes.

The basic wind turbine design consists of a bladed rotor that drives a shaft to a generator. The generator uses electromagnetic induction to produce a voltage. Wind power was used to enable electrolysis, the separation of water into hydrogen and oxygen, to make it possible to store energy as early as the 1890s.

Today, small-scale wind turbines for individual use generally have a maximum output of 400-1600 watts. In contrast, the largest industrial turbines might generate as much as 7.5 megawatts. Groupings of turbines are known as wind farms or wind parks.

Although windmills are still in use, wind turbines have become more common as the interest in renewable energy increases.

Biomass Energy

The term "biomass" refers to organic matter that has stored energy through the process of photosynthesis. It exists in one form as plants and may be transferred through the food chain to animals' bodies and their wastes, all of which can be converted for everyday human use through processes such as combustion, which releases the carbon dioxide stored in the plant material. Many of the biomass fuels used today come in the form of wood products, dried vegetation, crop residues, and aquatic plants. Biomass has become one of the most commonly used renewable sources of energy in the last two decades, second only to hydropower in the generation of electricity. It is such a widely utilized source of energy, probably due to its low cost and indigenous nature, that it accounts for almost 15% of the world's total energy supply and as much as 35% in developing countries, mostly for cooking and heating.

Biomass is one of the most plentiful and well-utilised sources of renewable energy in the world. Broadly speaking, it is organic material produced by the photosynthesis of light. The chemical material (organic compounds of carbons) are stored and can then be used to generate energy. The most common biomass used for energy is wood from trees. Wood has been used by humans for producing energy for heating and cooking for a very long time.

Biomass has been converted by partial-pyrolisis to charcoal for thousands of years. Charcoal, in turn has been used for forging metals and for light industry for millenia. Both wood and charcoal formed part of the backbone of the early Industrial Revolution (much northern England, Scotland and Ireland were deforested to produce charcoal) prior to the discovery of coal for energy.

Wood is still used extensively for energy in both household situations, and in industry, particularly in the timber, paper and pulp and other forestry-related industries. Woody biomass accounts for over 10% of the primary energy consumed in Austria, and it accounts for much more of the primary energy consumed in most of the developing world, primarily for cooking and space heating.

It is used to raise steam, which, in turn, is used as a by-product to generate electricity. Considerable research and development work is currently underway to develop smaller gasifiers that would produce electricity on a small-scale. For the moment, however, biomass is used for off-grid electricity generation, but almost exclusively on a large-, industrial-scale.

There are two issues that affect the evaluation of biomass as a viable solution to our energy problem: the effects of the farming and production of biomass and the effects of the factory conversion of biomass into usable energy or electricity. There are as many environmental and economic benefits as there are detriments to each issue, which presents a difficult challenge in evaluating the potential success of biomass as an alternative fuel. For instance, the replacement of coal by biomass could result in "a considerable reduction in net carbon dioxide emissions that contribute to the greenhouse effect." On the other hand, the use of wood and other plant material for fuel may mean deforestation. We are all aware of the problems associated with denuding forests, and widespread clear cutting can lead to groundwater contamination and irreversible erosion patterns that could literally change the structure of the world ecology.

Biomass has to be considered in the search for an alternative source of energy that is abundant in a wide-scale yet non-disruptive manner, since it is capable of being implemented at all levels of society. Although tree plantations have "considerable promise" in supplying an energy source, "actual commercial use of plantation-grown fuels for power generation is limited to a few isolated experiences." Supplying the United States ' current energy needs would require an area of one million square miles. That's roughly one-third of the area of the 48 contiguous states. There is no way that plantations could be implemented at this scale, not to mention that soil exhaustion would eventually occur. Biomass cannot replace our current dependence on coal, oil, and natural gas, but it can complement other renewables such as solar and wind energy.

According to Flavin and Lenssen of the Worldwatch Institute , "If the contribution of biomass to the world energy economy is to grow, technological innovations will be needed, so that biomass can be converted to usable energy in ways that are more efficient, less polluting, and at least as economical as today's practices." When we have enough government support and have allotted enough land for the continuous growth of energy crops for biomass-based energy, we may have a successful form of alternative energy. But "as long as worldwide prices of coal, oil and gas are relatively low, the establishment of plantations dedicated to supplying electric power or other higher forms of energy will occur only where financial subsidies or incentives exist or where other sources of energy are not available." Although it is currently utilized across the globe, biomass energy is clearly not capable of sustaining the world's energy needs on its own.

Hydropower

Hydropower or hydroelectricity refers to the conversion of energy from flowing water into electricity. It is considered a renewable energy source because the water cycle is constantly renewed by the sun.

Historically, one of the first uses of hydro power was for mechanical milling, such as grinding grains. Today, modern hydro plants produce electricity using turbines and generators, where mechanical energy is created when moving water spins rotors on a turbine. This turbine is connected to an electromagnetic generator, which produce electricity when the turbine spins.

Hydro plant facilities can be categorized into three sizes: large (>30 MW), small (100 kW - 30 MW), or micro (<100 kW).

There are Three Main Types of hydro Plants

1. Impoundment facilities are the most common technology which uses a dam to create a large reservoir of water. Electricity is made when water passes through turbines in the dam.
2. Pumped storage facilities are similar but have a second reservoir below the dam. Water can be pumped from the lower reservoir to the upper reservoir, storing energy for use at a later time.
3. Run-of-river facilities rely more on natural water flow rates, diverting just a portion of river water through turbines, sometimes without the use of a dam or reservoirs. Since run-of-river hydro is subject to natural water variability, it is more intermittent than dammed hydro.

Geothermal Energy

Geothermal energy is the energy that is harnessed from the Earth's internal heat and used for practical purposes, such as heating buildings or generating electricity. It also refers to the technology for converting geothermal energy into useable energy. The term geothermal power is used synonymously as the conversion of the Earth's internal heat into a useful form of energy, or more specifically as the generation of electricity from this thermal energy (geothermal electricity).

The four basic means for capturing geothermal energy for practical use are geothermal power plants (dry steam, flash steam, binary cycle), geothermal heat pumps, direct use, and enhanced geothermal systems.

Geothermal provides a huge, reliable, renewable resource, unaffected by changing weather conditions. It reduces reliance on fossil fuels and their inherent price unpredictability, and when managed with sensitivity to the site capacity, it is sustainable. Furthermore, technological advances have dramatically expanded the range and size of viable resources.

However, geothermal also faces challenges in the need for significant capital investment, and a significant amount of time in terms of building geothermal plants. There are limitations in terms of placement of geothermal plants in regions with accessible deposits of high temperature ground water, and construction of power plants can adversely affect land stability. Geothermal power plants also can lead to undesirable emissions, with power plant emitting low levels of carbon dioxide, nitric oxide, sulfur, methane, and hot water from geothermal sources may hold in solution trace amount of toxic elements, such as mercury, boron, and arsenic.

The Nesjavellir Geothermal Power Plant in Iceland.

Earth's Geothermal Resources

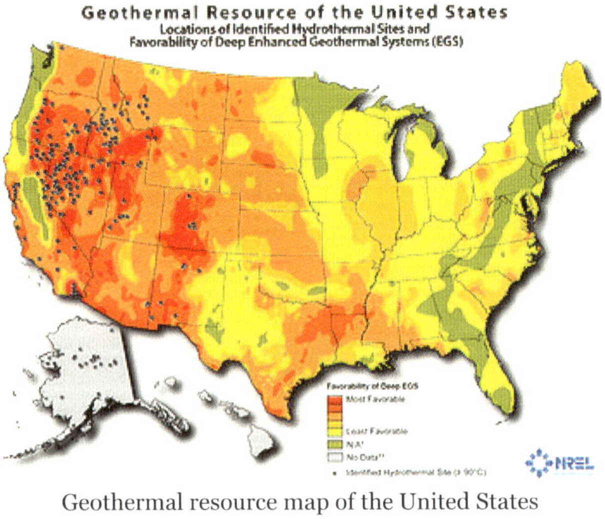

Geothermal resource map of the United States

The Earth's geothermal energy comes from the heat from the original formation of the planet (about 20%) and from the thermal energy continually generated by the radioactive decay of minerals (80%). The major heat-producing isotopes in Earth are potassium-40, uranium-238, uranium-235, and thorium-232 .

The Earth's internal thermal energy flows to the surface by conduction at a rate of 44.2 terawatts (TW), and is replenished by radioactive decay of minerals at a rate of 30 TW. These power rates are more than double humanity's current energy consumption from all primary sources, but most of this energy flow is not recoverable. In addition to the internal heat flows, the top layer of the surface to a depth of 10 meters (33 ft) is heated by solar energy during the summer, and releases that energy and cools during the winter.

Outside of the seasonal variations, the geothermal gradient of temperatures through the crust is 25–30 °C (77–86 °F) per kilometer of depth in most of the world. The conductive heat flux averages 0.1 MW/km^2. These values are much higher near tectonic plate boundaries where the crust is thinner. They may be further augmented by fluid circulation, either through magma conduits, hot springs, hydrothermal circulation or a combination of these.

Geothermal energy is considered "sustainable energy" and a "renewable energy resource" because the thermal energy is constantly replenished and the extraction by people is small relative to total content. Although the planet is slowly cooling, human extraction taps a minute fraction of the natural outflow, often without accelerating it.

The Earth's geothermal resources are theoretically more than adequate to supply humanity's energy needs, but only a very small fraction may be profitably exploited. Estimates of exploitable worldwide geothermal energy resources vary considerably. According to a 1999 study, it was thought that this might amount to between 65 and 138 GW of electrical generation capacity "using enhanced technology". This study did not assess the potential with significantly new technologies. Other estimates range from 35 to 2000 GW of electrical generation capacity, with a further potential for 140 EJ/year of direct use.

If heat recovered by ground source heat pumps is included, the non-electric generating capacity of geothermal energy is estimated at more than 100 GW (gigawatts of thermal power) and is used commercially in over 70 countries. A 2006 report by MIT that took into account the use of Enhanced Geothermal Systems (EGS) concluded that it would be affordable to generate 100 GWe (gigawatts of electricity) or more by 2050, just in the United States, for a maximum investment of 1 billion US dollars in research and development over 15 years (MIT 2006). The MIT report calculated the world's total EGS resources to be over 13 YJ, of which over 200 ZJ would be extractable, with the potential to increase this to over 2 YJ with technology improvements—sufficient to provide all the world's energy needs for several millennia.The total heat content of the Earth is 13,000,000 YJ.

Within about 10,000 meters (33,000 feet) of the Earth's surface there is considered to be about 50,000 times the amount of energy in geothermal energy resources as in all the world's oil and natural gas resources.

The world's biggest geothermal energy resources are in China; the second-largest ones in Hungary. By taking account of its size (about the size area of Illinois), Hungary has the richest such resources per sq mile/km. The world's biggest producer of electricity from geothermal sources is the Philippines. Other important countries are Nicaragua, Iceland, New Zealand.

Production

Ohaaki geothermal power station

Geothermal energy/power is produced by tapping into the thermal energy created and stored within the earth. The four basic categories for capturing geothermal energy for practical use are:

- Geothermal power plants (dry steam, flash steam, binary cycle)
- Geothermal heat pumps
- Direct use
- Enhanced geothermal systems

Geothermal energy is used commercially in over 70 countries (MIT 2006). In 2004, 200 petajoules (56 TWh) of electricity was generated from geothermal resources, and an additional 270 petajoules (75 TWh) of geothermal energy was used directly, mostly for space heating. In 2007, the world had a global capacity for 10 GW of electricity generation and an additional 28 GW of direct heating, including extraction by geothermal heat pumps. Heat pumps are small and widely distributed, so estimates of their total capacity are uncertain and range up to 100 GW.

Estimates of the potential for electricity generation from geothermal energy vary sixfold, from .035 to 2TW depending on the scale of investments. Upper estimates of geothermal resources assume enhanced geothermal wells as deep as 10 kilometers (6 mi), whereas existing geothermal wells are rarely more than 3 kilometers (2 mi) deep. Wells of this depth are now common in the petroleum industry.

In the United States, according to the Geothermal Energy Association's 2013 Annual GEA Industry Update, total installed U.S. geothermal capacity was estimated at 3,386 MW and the installed geothermal capacity grew by 5%, or 147.05 MW, since the previous annual survey in March 2012. This report noted that geothermal power plants were operating in eight states (Alaska, California, Hawaii, Idaho, Nevada, Oregon, Utah and Wyoming), and geothermal development was taking place in 6 more (Arizona, Colorado, North Dakota, New Mexico, Texas and Washington).

In the United States, most geothermal power plants are located in the western states (EIA 2011). California produces the most electricity from geothermal (EIA 2011), with installed capacity estimated to 2,732.2 MW in 2012, while the USA's second leading geothermal state, Nevada, reached 517.5 MW (GEA 2013). There are a number of geothermal plants concentrated in south central California, on the southeast side of the Salton Sea, near the cities of Niland and Calipatria, California. The Basin and Range geologic province in Nevada, southeastern Oregon, southwestern Idaho, Arizona, and western Utah is now an area of rapid geothermal development.

The type of source impacts which method can be used for capturing geothermal energy for production of electricity or other practical use. Flash plants are the most common way to generate electricity from liquid-dominated reservoirs (LDRs). LDRs are more common with temperatures greater than 200° C (392° F) and are found near young volcanoes surrounding the Pacific Ocean and in rift zones and hot spots. Pumps are generally not required, powered instead when the water turns to steam. Lower temperature LDRs (120-200° C) require pumping. They are common in extensional terrains, where heating takes place via deep circulation along faults, such as in the Western United States and Turkey. Lower temperature sources produce the energy equivalent of 100M BBL per year. Sources with temperatures from 30-150° C are used without conversion to electricity for such purposes as district heating, greenhouses, fisheries, mineral recovery, industrial process heating, and bathing in 75 countries. Heat pumps extract energy from shallow sources at 10-20° C for use in space heating and cooling. Home heating is the fastest-growing means of exploiting geothermal energy, with global annual growth rate of 30% in 2005, and 20% in 2012.

Heating is cost-effective at many more sites than electricity generation. At natural hot springs or geysers, water can be piped directly into radiators. In hot, dry ground, earth tubes or downhole heat exchangers can collect the heat. However, even in areas where the ground is colder than room temperature, heat can often be extracted with a geothermal heat pump more cost-effectively and cleanly than by conventional furnaces. These devices draw on much shallower and colder resources than traditional geothermal techniques. They frequently combine functions, including air conditioning, seasonal thermal energy storage, solar energy collection, and electric heating. Heat pumps can be used for space heating essentially anywhere.

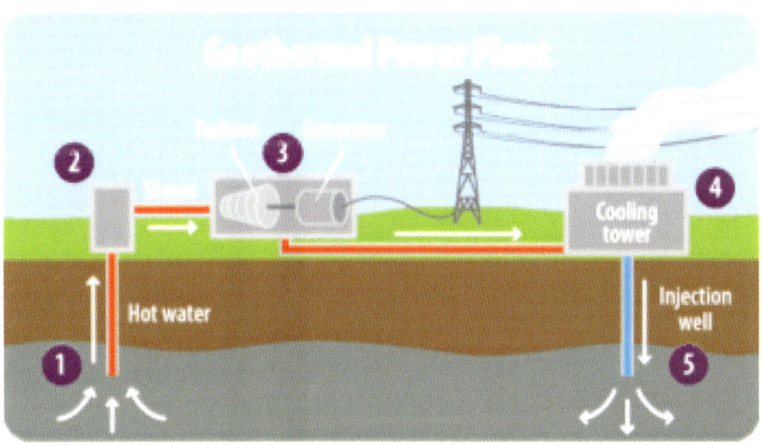

Generic geothermal power plant using hot water, with water pumped from deep underground under high pressure (1); water converts to steam when pressure drops at surface (2); steam spins a turbine connected to generator (3); steam cools off in cooling tower and becomes water (4); and cooled water pumped back into the earth.

Power Plants

Geothermal power plants use the heat from deep inside the Earth to pump hot water or hot steam to the surface to power generators. Such power plants drill their own wells into the rock to effectively capture the hot water or steam.

Such plants often are placed in places with plenty of geysers, active or geologically young volcanoes, or natural hot springs because these are areas where the Earth is particularly hot a reasonable distance from the surface. The water in such regions also can be more than 200°C (430°F) just below the surface.

There are three different designs for geothermal power plants: dry steam, flash steam, and binary cycle. These all bring hot water or steam from the ground, use it to power generators, and then the condensed steam and remaining geothermal fluid is injected back into the ground to pick up more heat and prolong the heat source. The design selected for generating power from geothermal energy depends on the temperature, depth, and quality of the water and steam in the area. If the hot water is high enough in temperature, flash system can be used. If it comes out as steam, it can be used directly to power the turbine with the dry stream design. If it is not high enough in temperature, then the binary cycle can be used to pass the water through a heat exchanger to heat up a second liquid that boils at a lower temperature than water and can be converted to steam to power the turbine.

Dry Steam

A dry stream power plant uses *hot steam*, typically above 235°C (455°F), to directly power its turbines. This is the oldest type of power plant and is still in use today. It is the simplest design in that steam goes directly through the turbine to power the generators, then is condensed into water in a cooling tower/condenser, and then returned to the ground.

The largest dry steam field in the world is The Geysers, 72 miles (116 km) north of San Francisco. The area was well known for hot springs, but actually does not have geysers, and the heat used is steam, not hot water. The Geysers began in 1960 and by 1990 had 26 power plants built in the area with a capacity of more than 2000 MW. However, the steam resource has been declining since 1988, due to the technology used and the rapid development of the area. The Geysers still had a net operating capacity of 725 MW by 2009 and the rocks underground remain hot.

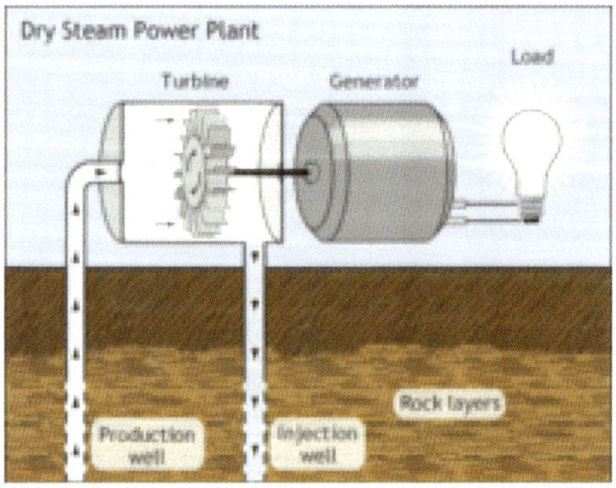

Basic dry steam power plant design

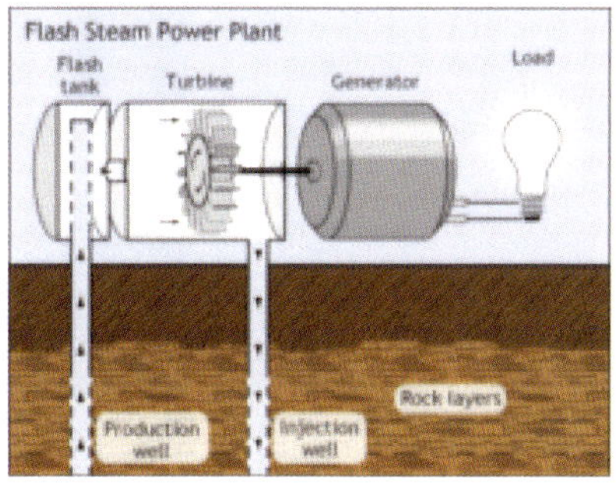

Basic flash steam power plant design

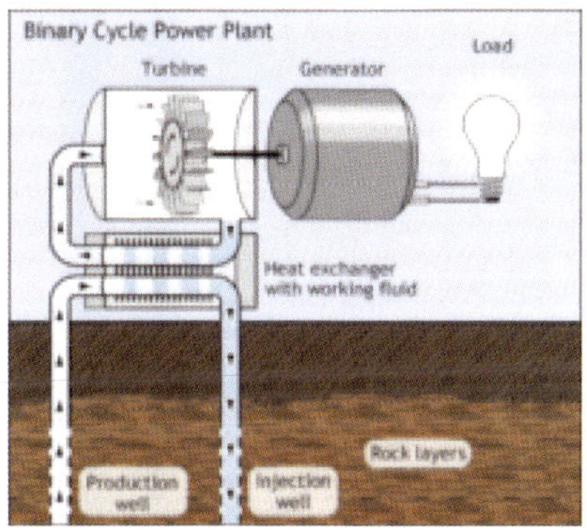

Basic binary cycle power plant design

Flash Steam

Flash steam power plants use *hot water* above 182°C (360°F) from geothermal reservoirs and has the addition of a flash tank over the dry steam design. As the water is pumped from the reservoir to the power plant, the drop in pressure in the flash tank causes the water to vaporize into steam (depressurized or "flashed" into steam), which then flows past the turbine, powering the electric generators. Any water not flashed into steam is injected back into the reservoir for reuse, as is the water that is captured from the steam after it has moved the turbines.

As noted above, flash steam plants are the most common way to generate electricity from liquid-dominated reservoirs (LDRs), which are often found near young volcanoes surrounding the Pacific Ocean and in rift zones and hot spots.

Binary Cycle

The third design, the binary cycle system or binary system, adds a heat exchanger as part of the design, in order to use *hot water* that is *cooler* than that of the flash steam plants. The hot fluid from geothermal reservoirs is passed through a heat exchanger, which transfers heat to a separate pipe containing fluids with a much lower boiling point, and thus more easily converted to steam. These fluids, usually Isobutane or Isopentane, running through a closed loop, are vaporized to produce the steam to power the turbine. The water from the ground is only used to transfer its heat to the second fluid and is returned to the ground.

The advantage to binary cycle power plants is their lower cost and increased efficiency. These plants also do not emit any excess gas and are able to utilize lower temperature reservoirs, which are much more common. Most geothermal power plants planned for construction are binary cycle.

Heat Pumps

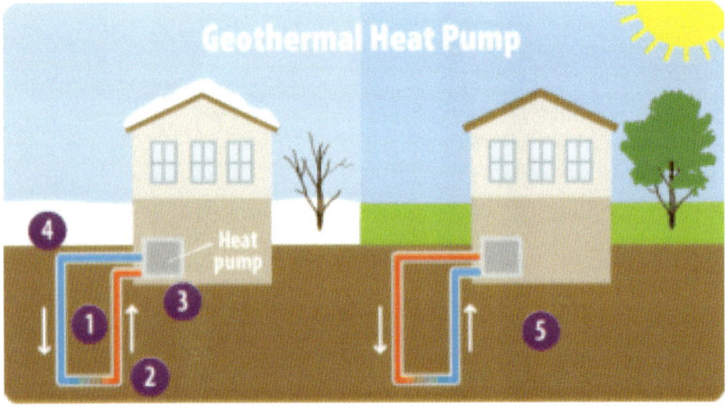

Generic geothermal heat pump, where water or refrigerant moves through loop of pipes (1); in times of cold weather, water or refrigerant heats up as transverses the part of the loop buried underground (2); warmed water or refrigerant transfers heat to building (3); now cooler water, after heat transferred to building, is pumped back underground to star the process again (4); and on hot days, the system can run in reverse (5).

A geothermal heat pump (GHP) can be used to extract heat from the ground to provide heat and cooling for buildings. Geothermal heat pumps are also known as ground-source heat pumps, Geo-Exchange heat pumps, earth-coupled heat pumps, and water-source heat pumps. These systems take advantage of the fact that a few feet below the Earth's surface, the temperature of the ground remains relatively constant and thus warmer than the air temperature in cold weather and colder than the air in warm weather. Using water or refrigerant, the pumps utilize pipes buried underground to move heat from the ground to the building during cold weather and from the building to the ground during warm weather. Some combine a air-source heat pump with a geothermal heat pump.

Heat pumps can range from simple systems involving a tube that runs from the outside air, under the ground, and then into a house's ventilation system. More complex systems involve compressors and pumps to maximize heat transfer. Enough heat can be extracted from shallow ground anywhere in the world to provide home heating, but industrial applications need the higher temperatures of deep resources.

GHPs can be much more efficient than electric heating and cooling, and are particularly energy-efficient in regions with temperature extremes. As of 2009, in the United States where were more than 600,000 geothermal heat pumps in use in homes and other buildings, with new installations at about 60,000 per year. The United States Department of Energy estimated that pumps can save a typical home hundreds of dollars in energy costs per year. However, GHPs have high up-front costs and installation can be difficult as it involves digging up areas around the building.

Four basic designs are typically utilized for geothermal heat pump systems: Horizontal closed loop systems, vertical closed loop systems, pond/lake closed loop systems, and the open-loop option. The are variants of these systems as well as hybrid systems that use different geothermal resources.

In general, closed loop systems typically circulate an antifreeze solution through a closed loop buried in the ground or immersed in water. Heat is transferred between the refrigerant in the heat pump and the antifreeze solution in the closed loop via a heat exchanger. The possible configurations for the loop are horizontal, vertical, and pond/lake. One variant, direct exchange, does not use a heat exchanger but instead pumps the refrigerant directly through tubing buried in the ground.

Open loop systems utilizes surface body water or well water as the heat exchange fluid and circulates this directly through the GPH system. After the water circulates through the system, it is returned to the ground through the well, a recharge well, or surface discharge. This requires a sufficient supply of relatively clean water.

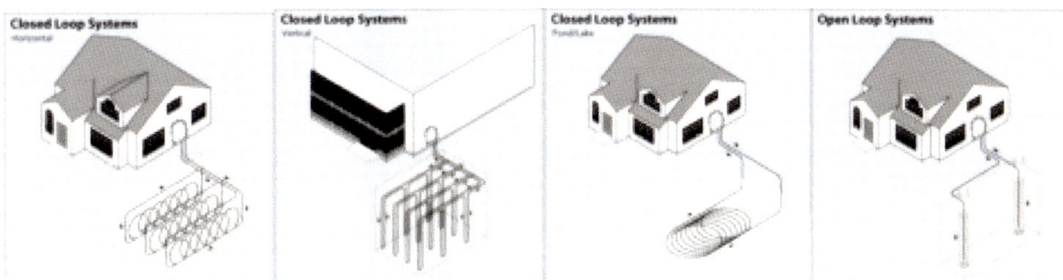

Direct use

Some areas have geothermal resources that can be used directly for heating purposes. For example, hot spring water is used for heating greenhouses, heat spas, heat fish farms, and so forth.

Iceland is the world leader in direct applications. More than fifty percent of its energy comes from geothermal resources (UCS 2009) and some 93% of its homes are heated with geothermal energy, saving Iceland over $100 million annually in avoided oil imports. Reykjavík, Iceland has the world's biggest district heating system, bringing in hot water from 25 kilometers way (UCS 2009). Once known as the most polluted city in the world, it is now one of the cleanest.

In the United States, Boise, Idaho and Klamath Falls, Oregon have used geothermal water to heat buildings and homes for more than a century.

Enhanced Geothermal Systems

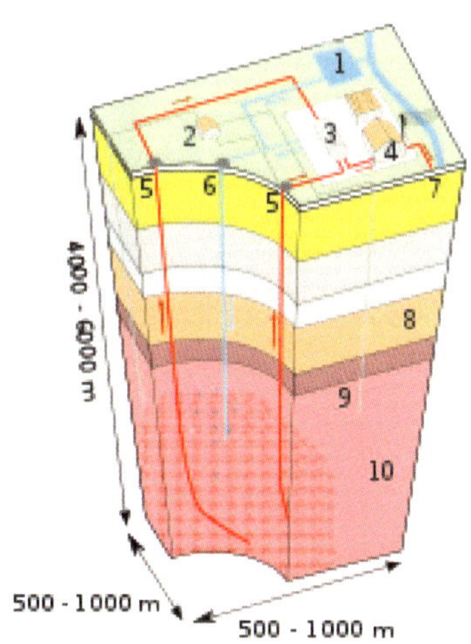

Enhanced geothermal system 1:Reservoir 2:Pump house 3:Heat exchanger 4:Turbine hall 5:Production well 6:Injection well 7:Hot water to district heating 8:Porous sediments 9:Observation well 10:Crystalline bedrock

Although geothermal heat is everywhere below the Earth's surface, only about ten percent of the land surface area has conditions where the water circulates near the surface to be easily captured (UCS 2009). Enhanced geothermal systems allow the capturing of heat even in these dry locations. It also is effective in capturing heat from locations where the natural supply of water producing steam from the hot underground magma deposits has been exhausted.

Enhanced geothermal systems (EGS) actively inject water into wells to be heated and pumped back out. The water is injected under high pressure to expand existing rock fissures to enable the water to freely flow in and out. The technique was adapted from oil and gas extraction techniques. However, the geologic formations are deeper and no toxic chemicals are used, reducing the possibility of environmental damage. Drillers can employ directional drilling to expand the size of the reservoir.

The key characteristic of an EGS is that it reaches at least 10 km down into hard rock. Drilling at this depth is now routine for the oil industry. Exxon announced an 11 km hole at the Chayvo field, Sakhalin. At a typical EGS site two holes would be bored and the deep rock between them fractured. Water would be pumped down one and steam would come up the other. The technological challenges are to drill wider bores and to break rock over larger volumes. Apart from the energy used to make the bores, the process releases no greenhouse gases.

The world's total EGS resources have been estimated to be over 13,000 ZJ, of which over 200 ZJ would be extractable, with the potential to increase this to over 2,000 ZJ with technology improvements—sufficient to provide all the world's energy needs for 30,000 years.

Electricity

The International Geothermal Association (IGA) reported in 2010 that 10,715 megawatts (MW) of geothermal power in 24 countries was online and was expected to generate 67,246 GWh of electricity in 2010. This represents a 20% increase in online capacity since 2005.

In 2010, the United States led the world in geothermal electricity production with 3,086 MW of installed capacity from 77 power plants (Holm et al. 2010). The largest group of geothermal power plants in the world is located at The Geysers. The Philippines is the second highest producer, with 1,904 MW of capacity online in 2010; Geothermal power makes up approximately 27% of Philippine electricity generation.

Geothermal power is generated in over 20 countries around the world including Iceland (producing over 26 percent of its electricity from geothermal sources in 2006), the United States, Italy, France, New Zealand, Mexico, Nicaragua, Costa Rica, Russia, the Philippines (production capacity of 1931 MW (2nd to US, 27 percent of electricity), Indonesia, the People's Republic of China, and Japan. Canada's government (which officially notes some 30,000 earth-heat installations for providing space heating to Canadian residential and commercial buildings) reports a test geothermal-electrical site in the Meager Mountain–Pebble Creek area of British Columbia, where a 100 MW facility could be developed.

Installed geothermal electric capacity				
Country	Capacity (MW)	Capacity (MW)2010	Percentage of national electricity production	Percentage of global geothermal production
United States	2687	3086	0.3	29
Philippines	1969.7	1904	27	18
Indonesia	992	1197	3.7	11
Mexico	953	958	3	9
Italy	810.5	843	1.5	8
New Zealand	471.6	628	10	6
Iceland	421.2	575	30	5
Japan	535.2	536	0.1	5
Iran	250	250		
El Salvador	204.2	204	25	
Kenya	128.8	167	11.2	
Costa Rica	162.5	166	14	
Nicaragua	87.4	88	10	
Russia	79	82		
Turkey	38	82		
Papua-New Guinea	56	56		
Guatemala	53	52		
Portugal	23	29		
China	27.8	24		
France	14.7	16		
Ethiopia	7.3	7.3		
Germany	8.4	6.6		
Austria	1.1	1.4		
Australia	0.2	1.1		
Thailand	0.3	0.3		
TOTAL	9,981.9	10,959.7		

In the United States, geothermal is one of the renewable energy resources used to produce electricity, but its growth is slower than that of wind and solar energy development and a November 2011 report noted that it produced just 0.4% of the electricity from all sectors nationally during the first 8 months of that year, with 10,898 million kilowatthours (kWh) produced during that time. However, about 5% of the electricity generated in California was produced from geothermal, although there are significant geothermal resources that could be utilized.

Geothermal thermal energy is used to generate electricity typically via a well that is drilled into an underground reservoir of water that can be as hot as 371 degrees Celsius (700 Fahrenheit). Geothermal electric plants were traditionally built exclusively on the edges of tectonic plates where high temperature geothermal resources are available near the surface. The development of binary cycle power plants and improvements in drilling and extraction technology enable enhanced geothermal systems over a much greater geographical range.

The thermal efficiency of geothermal electric plants is low, around 10–23%, because geothermal fluids do not reach the high temperatures of steam from boilers. The laws of thermodynamics limits the efficiency of heat engines in extracting useful energy. Exhaust heat is wasted, unless it can be used directly and locally, for example in greenhouses, timber mills, and district heating. System efficiency does not materially affect operational costs as it would for plants that use fuel, but it does affect return on the capital used to build the plant. In order to produce more energy than the pumps consume, electricity generation requires relatively hot fields and specialized heat cycles. Because geothermal power does not rely on variable sources of energy, unlike, for example, wind or solar, its capacity factor can be quite large—up to 96% has been demonstrated.

Benefits and Challenges

Geothermal energy offers a huge, reliable, renewable resource. It is sustainable when managed with sensitivity to the site capacity; for example, the hot water extracted in the geothermal process can be re-injected into the ground to produce more steam. It also is a source that is unaffected by changing weather conditions. Furthermore, technological advances have dramatically expanded the range and size of viable resources, especially for applications such as home heating, opening a potential for widespread exploitation. Geothermal wells do release greenhouse gases trapped deep within the earth, but these emissions are much lower per energy unit than those of fossil fuels.

From an economic view, geothermal energy is price competitive in some areas. It also reduces reliance on fossil fuels and their inherent price unpredictability; geothermal power requires little fuel, except for purposes like pumps. Given enough excess capacity, geothermal energy can also be sold to outside sources such as neighboring countries or private businesses that require energy. It also offers a degree of scalability: a large geothermal plant can power entire cities while smaller power plants can supply more remote sites such as rural villages.

Geothermal has minimal land and freshwater requirements. Geothermal plants use 3.5 square kilometers (1.4 sq mi) per gigawatt of electrical production (not capacity) versus 32 square kilometers (12 sq mi) and 12 square kilometers (4.6 sq mi) for coal facilities and wind farms respectively (Lund 2007). They use 20 liters of freshwater per MW·h versus over 1000 liters per MW·h for nuclear, coal, or oil.

Several entities, such as the National Renewable Energy Laboratory and Sandia National Laboratories, conduct research toward the goal of establishing a proven science around geothermal energy. The International Centre for Geothermal Research (IGC), a German geosciences research organization, is largely focused on geothermal energy development research.

Krafla Geothermal Station in northeast Iceland

Geothermal power station in the Philippines

However, use of geothermal energy also faces several challenges. For one, geothermal plants generally are site-specific and limited to regions with accessible deposits of high temperature ground water. Capital costs also are significant. Drilling and exploration for deep resources is very expensive. Drilling accounts for over half the costs, and exploration of deep resources entails significant risks. The completing of a geothermal plant takes significant time (four to eight years) versus the times for wind or solar, and there is a lack of transmission lines.

There also are several environmental concerns behind geothermal energy.

For one, there can be negative impacts on surrounding lands. Construction of the power plants can adversely affect land stability in the surrounding region and land subsidence can become a problem as older wells begin to cool down. Also, increased seismic activity can occur because of well drilling. Subsidence has occurred in the Wairakei field in New Zealand. In Staufen im Breisgau, Germany, tectonic uplift occurred instead, due to a previously isolated anhydrite layer coming in contact with water and turning into gypsum, doubling its volume. Enhanced geothermal systems can trigger earthquakes as part of hydraulic fracturing. The project in Basel, Switzerland was suspended because more than 10,000 seismic events measuring up to 3.4 on the Richter Scale occurred over the first 6 days of water injection.

Geothermal power plants can also lead to undesirable emissions. Dry steam and flash steam power plant emit low levels of carbon dioxide, nitric oxide, and sulfur, although at roughly 5 percent of the levels emitted by fossil fuel power plants. Fluids drawn from the deep earth carry a mixture of gases, notably carbon dioxide (CO_2), hydrogen sulfide (H_2S), methane (CH_4) and ammonia (NH_3). These pollutants contribute to acid rain, and noxious smells if released, and include some important greenhouse gases. Existing geothermal electric plants emit an average of 122 kilograms (270 lb) of CO_2 per megawatt-hour (MW·h) of electricity, a small fraction of the emission intensity of conventional fossil fuel plants. Plants that experience high levels of acids and volatile chemicals are usually equipped with emission-control systems to reduce the exhaust.

In addition to dissolved gases, hot water from geothermal sources may hold in solution trace amounts of toxic elements such as mercury, arsenic, boron, and antimony. These chemicals precipitate as the water cools, and can cause environmental damage if released. The modern practice of injecting cooled geothermal fluids back into the Earth to stimulate production has the side benefit of reducing this environmental risk.

Direct geothermal heating systems contain pumps and compressors, which may consume energy from a polluting source. This parasitic load is normally a fraction of the heat output, so it is always less polluting than electric heating. However, if the electricity is produced by burning fossil fuels, then the net emissions of geothermal heating may be comparable to directly burning the fuel for heat. For example, a geothermal heat pump powered by electricity from a combined cycle natural gas plant would produce about as much pollution as a natural gas condensing furnace of the same size. Therefore the environmental value of direct geothermal heating applications is highly dependent on the emissions intensity of the neighboring electric grid.

Solar Energy

Solar energy is a kind of renewable energy provided by the sun.

This energy is in the form of solar radiation, which makes the production of solar electricity possible.

Electricity can be produced directly from photovoltaic, PV, cells. (Photovoltaic literally means "light" and "electric.") These cells are made from materials which exhibit the "photovoltaic effect" i.e. when sunshine hits the PV cell, the photons of light excite the electrons in the cell and cause them to flow, generating electricity.

Solar energy produces electricity when it is in demand during the day particularly hot days when air-conditioners drive up electricity demand.

In use, solar energy produces no emissions. One megawatt hour of solar electricity offsets about 0.75 to 1 tonne of CO_2.

PV panels are being used increasingly, both in the city and in remote locations, to produce electricity for households, schools and communities, and to supply power for equipment such as telecommunication and water pumps. The majority of solar PV installations in Australia are grid-connected systems.

Also, electricity for remote and regional Australian communities has been supplied by solar energy for many years.

Australia is one of the sunniest countries in the world and there is huge potential for solar PV to make a significant contribution to electricity generation.

References

- Renewable-energy: newworldencyclopedia.org, Retrieved 09 July 2018
- Wind-power: whatis.techtarget.com, Retrieved 30 June 2018
- Hydro-power: studentenergy.org, Retrieved 14 July 2018
- Geothermal-energy: newworldencyclopedia.org, Retrieved 28 April 2018
- What-solar-energy, solar-energy: greenoughsolarfarm.com.au, Retrieved 10 March 2018

8

Energy Economics and Efficiency

The study of the supply and use of energy in a society falls under the discipline of energy economics. In an energy conversion machine, the ratio between the useful generated output and the input is known as energy efficiency. The following chapter elucidates the fundamentals of energy efficiency gap, fuel and fuel efficiency, energy efficiency in transport, etc. as well as the principles of energy economics for a close understanding of energy economics and efficiency.

Energy Economics

Energy economics is the field that studies human utilization of energy resources and energy commodities and the consequences of that utilization. In physical science terminology, "energy" is the capacity for doing work, e.g., lifting, accelerating, or heating material. In economic terminology, "energy" includes all energy commodities and energy resources, commodities or resources that embody significant amounts of physical energy and thus offer the ability to perform work. Energy commodities - e.g., gasoline, diesel fuel, natural gas, propane, coal, or electricity – can be used to provide energy services for human activities, such as lighting, space heating, water heating, cooking, motive power, electronic activity. Energy resources - e.g., crude oil, natural gas, coal, biomass, hydro, uranium, wind, sunlight, or geothermal deposits – can be harvested to produce energy commodities.

Energy economics studies forces that lead economic agents – firms, individuals, governments – to supply energy resources, to convert those resources into other useful energy forms, to transport them to the users, to use them, and to dispose of the residuals. It studies roles of alternative market and regulatory structures on these activities, economic distributional impacts, and environmental consequences. It studies economically efficient provision and use of energy commodities and resources and factors that lead away from economic efficiency.

Properties of Energy Resources and Energy Commodities

Other than all embodying significant amounts of physical energy, energy resources or commodities vary greatly. They may embody chemical energy (e.g., oil, natural gas, coal, and biomass), mechanical energy (e.g., wind, falling water), thermal energy (geothermal deposits), radiation (sunlight, infrared radiation), electrical energy (electricity), or the potential to create energy through nuclear reactions (uranium, plutonium.) They have differing physical forms. Crude oil, most refined petroleum products, and water are liquids. Of water includes available energy only through its motion. Coal, most biomass, and uranium are solids. Natural gas and wind are in gases, with wind including available energy based only on its movement. Geothermal energy is available through hot liquids (normally water) or solids (subterranean rock formations). Solar radiation is a pure form of energy. Electricity consists of electrons moving under an electrical potential.

Resources can be viewed as renewable or depletable. Some renewable resources can be stored; others are not storable.

Energy Conversion Processes

A fundamental property of energy is expressed by the first law of thermodynamics: energy can be neither created nor destroyed (except through nuclear reactions transforming matter to energy.) Energy can be converted between forms and human use of energy typically involves such conversions for human ends.

Energy conversion processes are basic to human experience. Fire, providing heat and light, is a process by which chemical energy stored in the fuel, say, wood, is converted to thermal energy and radiant energy. Chemical energy stored in wood is the result of photosynthesis, whereby plants convert energy in sunlight to chemical energy, stored in the plant material. Carbohydrates in food are converted within the human body to thermal energy and mechanical energy, providing body warmth and movement.

The industrial revolution was characterized by a change from use of hand tools, using human mechanical energy, to machine and power tools. Machine tools allowed conversion of energy in falling water to mechanical energy (water wheels) or conversion of chemical energy in wood or coal to mechanical energy (steam engines) for industrial processes.

Humans now routinely harness complex sequences of energy conversion processes to provide desired services. Crude oil is separated into refined products such as gasoline, diesel oil, jet fuel, heavy distillates that embody chemical energy. Gasoline or diesel oil are explosively burned in internal combustion engines, converting chemical energy into thermal energy. Heated gases push engine pistons, converting thermal energy into mechanical energy. Some is lost as heated gases or as radiant energy. The mechanical energy moves the automobile and, in the process, is converted to thermal energy through friction within the automobile or between the automobile and the road or air. Some mechanical energy is converted to electrical energy by a generator in the automobile, to power electrical equipment. Some electrical energy is converted into chemical energy in the automobile battery. To start the car, chemical energy in the battery is converted to electrical energy, that is then converted to mechanical energy to turn the engine.

Similarly, coal combustion converts chemical energy into thermal energy to create steam, which powers a turbine in an electric generating plant, converting the thermal energy into mechanical energy and then into electrical energy. Electricity can power a motor (converting to mechanical energy), heat a room (to thermal energy), or light a bulb (to thermal energy, then to light). This energy is later converted to thermal energy, which ultimately is radiated into space.

Energy economics recognizes the fundamental physical realities that 1) no energy is created or destroyed but that energy can be converted among its various forms, and 2) energy comes from the physical environment and ultimately is released back into the physical environment. Thus, energy economics is the study of human activities using energy resources from naturally available forms, through often complex conversion processes, to forms providing energy services.

Demand for Energy as a Derived Demand

Demand for energy is derived from wishes to use energy to obtain desired services. It is not derived from preferences for the energy commodity itself. Energy demand depends primarily on demand for desired services, availability and properties of energy conversion technologies, and costs of energy and technologies used for conversion.

For example, consumers use gasoline to fuel an automobile or other motorized vehicle, converting gasoline to mechanical energy for motive power. The amount of gasoline used is proportional to the miles the auto is driven and inversely proportionate to the efficiency by which gasoline is converted to useful mechanical energy, measured as miles per gallon (Mpg) of gasoline of the automobile. Demand for gasoline is thus derived from choices about distances vehicles are driven and their energy conversion efficiencies.

Similarly, electricity is purchased by consumers only to perform functions using electricity. Typical electricity uses include lighting, refrigeration, space heating, air conditioning, clothes washing, drying, dish washing, water heating, operating electronic equipment such as computers or televisions. Electrical energy is converted to mechanical energy (motors in refrigerators, air-conditioning units, vacuum cleaners), thermal energy (space heating, clothes dryers, water heating), or radiation (lighting, television, computer monitors.) Electricity demand is derived from demand for the underlying services – comfortable space, refrigeration, cleaning, entertainment, information processing.

In each case, efficiency of energy conversion equipment also determines energy demand. Typically, energy conversion equipment is long-lived – automobiles, air-conditioning units, refrigerators, televisions, computer systems, furnaces. Consumers or firms can usually choose among alternatives with various conversion efficiencies; such choices significantly influence energy demand. To the extent that consumers and firms purchase these units with an understanding of their conversion efficiencies, expectations of future energy prices can influence choices of particular equipment. For example, high natural gas prices can motivate consumers to invest in home insulation.

In general, increased energy prices reduce demand by reducing use of energy services and motivating selection of higher conversion efficiency equipment. For example, gasoline prices influence demand through vehicle miles and fuel efficiency of vehicles. Vehicle miles is influenced by cost per mile of driving, including per mile gasoline costs, equal to the ratio Pg/Mpg (where Pg is the gasoline price), and other costs. Increased gasoline prices lead consumers to purchase more fuel efficient cars. Both factors imply that increased gasoline prices reduce gasoline demand, with the vehicle miles adjusting relatively quickly and vehicular fuel efficiency adjusting slowly as vehicles enter the fleet.

Except for firms selling energy resources or energy commodities, the same issues are important for industrial and commercial use of energy.

Demand Substitution among Energy Commodities

Some energy services can be provided by several different energy commodities. Homes could be heated using electricity, natural gas, oil, or wood, since each can be converted to thermal energy. Cooking could use electricity, natural gas, propane, wood, or charcoal. Thus, energy commodities are typically economic substitutes for one another: the demand for a particular energy commodity is an increasing function of prices of other energy commodities.

This substitutability of energy is made possible by and is limited by the available set of energy conversion technologies. Typically one conversion technology can be used only for one particular energy commodity. For home heating, a natural gas furnace cannot use oil, electricity, or wood. Because conversion equipment typically is very long lived, substitution among energy commodities occurs only slowly, and then when new equipment is purchased. Short-run substitution usually can occur only if several energy conversion technologies are simultaneously available for use by particular consumers, e.g., homes that have a central natural gas heating system plus portable electric space heating units. Thus, usually various energy commodities can be viewed as imperfect substitutes for one another, with much greater substitutability in the long run than in the short run.

The degree of substitutability can be sharply altered by development of new conversion alternatives. For example, automobiles historically were fueled only by gasoline or diesel fuel, but technologies currently being developed would allow autos to be powered by electricity, natural gas, propane, hydrogen, or other energy commodities. Once such conversion technologies are successfully commercialized, gasoline and other energy commodities will become highly substitutable in transportation.

Energy an Essential Good

In economics, an essential good is one for which the demand remains positive no matter how high its price becomes. In the theoretical limit, for prices unboundedly high, consumers would allocate all of their income to purchases of the essential good.

Energy is often described as an essential good because human activity would be impossible absent use of energy: living requires food embodying chemical energy. Although energy is essential to humans, neither particular energy commodities nor any purchased energy commodities are essential goods. Particular energy commodities are not essential because consumers can convert one form of energy into another. Even the aggregate of all purchased energy cannot be viewed as an essential good. Experience from low-energy research facilities shows that an extremely energy efficient home needs relatively little energy. Solar energy could generate electricity or heat water. Travel could be limited to walking or riding bicycles. Solar-generated electricity or wood fires could be used for cooking. For high enough prices of purchased energy, demand for purchased energy by consumers could be reduced to zero. Thus, purchased energy is not an essential good.

Optimality of Consumer Choice

Debate is ongoing about the extent to which consumers understand conversion efficiencies of alternative technologies and act on this understanding and the extent to which manufacturers of conversion equipment respond to consumer preferences. Labeling requirements have been one policy response to concerns that consumers otherwise would have insufficient information to choose among energy conversion equipment. New cars have stickers with estimates of their Mpg under different driving cycles. Refrigerator labels estimate their annual electricity use and cost. The concern has led to adoption of energy efficiency standards for household equipment, such as refrigerators, under the belief that labeling is not sufficient to motivate optimal consumer choices. Similarly, the imposition of corporate average fuel efficiency standards (CAFE) for automobiles is a legislative response to concern that, on their own, automobile manufacturers will make automobiles less fuel efficient than optimal. These behavioral issues and the appropriate policy responses have not been fully resolved in either the existing literature or the policy community.

Depletable, Storable Renewable, Non-storable Renewable Resources

Based on the speed of natural processes, one can classify primary energy resources as depletable or renewable. Renewable resources can be further subdivided into storable or nonstorable resources. Renewable resources are self-renewing within a time scale important for economic decision making. Storable renewable resources typically exist as a stock which can be used or can be stored. Biomass, hydro power, and some geothermal, fall in this category. The amount used at one time influences the amount available in subsequent times. Nonstorable renewable resources – wind, solar radiation, run-of-the-river hydro resources can be used or not, but the quantity used at a given time has no direct influence on the quantity available subsequently. Most energy commodities are storable (refined petroleum products, processed natural gas, coal, batteries), but electricity is not storable as electricity. Depletable resources are those whose renewal speeds are so slow that it is appropriate to view them as made available once and only once by nature. Crude oil, natural gas, coal, and uranium all fall in this category.

Initially all human energy use depended on renewable resources, in particular biomass resources used for food, heat, or light.

Depletable Resource Economics and the Transition to Renewable Resources

The study of depletable resource economics began with articles by Lewis Grey and Harold Hotelling, which examined economically inter-temporal optimal extraction from a perfectly known stock of the resource, with perfectly predictable future prices of the extracted commodity. Many, but not all, subsequent articles maintained the perfect knowledge assumptions.

The essential result is that under optimal extraction paths the resource owner recognizes (explicitly or implicitly) an opportunity cost, or rent, in addition to the marginal extraction costs. All information about the role of future prices and costs would be embodied in this opportunity cost. The competitive firm would extract at a rate such that the marginal extraction cost plus opportunity cost would equal the selling price for the extracted commodity. Price would thus exceed marginal cost, even if the firm were operating perfectly competitively. This opportunity cost would evolve smoothly over time. As the resource neared depletion, the opportunity cost and the marginal extraction cost even at very low extraction rates would together have increased until they equaled the commodity price, at which time extraction would cease.

In depletable resource theory, market prices would increase gradually to the cost of producing substitutes, reaching that cost only as the depletable resource were nearing depletion.

Substitutes would be produced only in small quantities until near the time of depletion. Market forces would automatically and optimally guide commodity prices upward so that when the depletable resources were nearing depletion, commodity prices would have risen to a point at which the demands could be fully satisfied by the substitutes.

In reality, the economic cycle for depletable resources is far more complex and more prone to error and surprises. The cycle typically begins with innovations that allow the resource to be utilized.

Technologies improve over time, partially guided by economic forces and public policy decisions, but often in somewhat unpredictable ways. Generally the magnitude and location of the resource base remains unknown and exploration is required to identify resource deposits. But exploration is costly. Therefore typically it is optimal for companies to explore only until they find sufficient resources to satisfy their expectations of near-term extraction. These discovered resources – referred to as "reserves" – are typically only a fraction of the resource base and do not provide reliable estimates of the overall size of the resource stock. Firms may optimally extract from the proven reserves but cannot know with any certainty the quantity or extraction costs of the undiscovered resources. Opportunity cost would depend on the future prices of the extracted commodity, but future commodity prices will themselves depend on future demand and supply, which in turn depends on the uncertain future discoveries, and which therefore are very unpredictable.

With this additional complexity and uncertainty, although the opportunity cost concept remains important, it seems appropriate to focus more attention on responses of markets to random changes in technologies, reserves, and other market information. And it seems appropriate to abandon the notion that markets will automatically and optimally guide the system to a smooth transition to renewable resources.

One central result does remain, however. If there is only a limited stock of a resource – undiscovered plus discovered – then there will be only a limited number of years during which the resource can be extracted. Here there is an important commonality between energy economics and ecological economics. The typical pattern includes an initial period in which the resource is not used, before technology for extracting and/or using the resource is developed. Extraction rates rise over time, perhaps rapidly, as that technology develops and demand increases. Commodity prices would fall with falling extraction and finding costs. However, at some time, rising costs due to depletion of the resource start overtaking the decreasing costs due to technology advances. The extraction rate declines until ultimately all of the economical resource stocks are depleted. At that point, the consumers of the depletable resource must substitute some other means of satisfying energy service demand.

If markets work well, the renewable resources will then be available in sufficient quantities and at reasonable costs. A transition to renewable energy resources will have been successfully accomplished and there would always be an adequate energy supply to satisfy all demands at the prevailing market price.

However, given the complexity and the uncertainty, it is not obvious that the transition will work as automatically, as smoothly, or as optimally as suggested above. It may be that fossil fuels saved for the future will ultimately never be needed because substitute forms of energy become available at a lower cost and at an earlier time than expected. Or it may be that fossil fuels are rapidly depleted but that costs of renewable resources remain well above expected levels or that quantities of renewable resources are more limited than expected. In either case, in retrospect everyone might wish to have made very different public policy decisions.

Energy Conversion Industries

In addition are activities associated with commercial conversion of energy from one form to another, particularly to electricity, from hydro power, coal, natural gas, oil, nuclear fission, wood and waste products, geothermal, wind, or solar radiation.

Energy conversion industries, for economic success, must be able to sell their product at a price higher than the cost of energy commodities used as inputs plus per unit capital and operating costs of the facilities. Energy conversion is never perfectly efficient and some input energy is lost into the environment. Therefore, the price per Btu of electricity must be substantially greater than the price of energy commodities used to generate electricity.

Technological advance can be very important. New technologies are becoming available that increase the conversion efficiency from natural gas or coal to electricity and which can be expected to have lower operating and capital costs. Such technological advances can be expected to bring prices of these energy commodities closer together over time.

In addition to these technological changes, there are important ongoing changes in economic structure of the electricity production and distribution industry, throughout the world.

In many countries, state-owned industries generate, transmit, and distribute electricity. In others, private electricity suppliers are subject to special economic regulation. The reason for governmental ownership or control seems to stem from two factors. Electricity is fundamental to economic activity and many people have not trusted private industry. Second, production, transmission, and distribution of electricity have shown significant increasing returns to scale and the industry has been viewed as a natural monopoly. Fearful that an unregulated monopoly would exercise market power and overprice electricity, most nations have chosen to tightly control or own the industry.

Recently, however, smaller geographically distributed electric generating plants, that could reasonably compete with one another, have become economically attractive. Thus, the possibility for competition in electricity generation has been recognized. In addition, it is now realized that an electric utility sells two classes of products: electricity delivery services (wires) and electricity. Although these two classes of products traditionally were bundled together into a price per kilowatt hour of electricity, in principle, these two classes could be unbundled and sold by separate companies. Electricity delivery service is characterized by increasing returns to scale, but electricity itself is not. Therefore the possibility is open for a competitive market structure to sell electricity to consumers, separately from the electricity delivery services.

In some localities this movement toward privatization and deregulation seems to be very successful; in others, for example, California, it has not been. At the time of this writing, there is an intense heated debate about whether deregulation of the electricity industry is appropriate and if so, what is the appropriate form of deregulation.

Environmental Consequences of Energy Use

Many important environmental damages stem from the production, conversion, and consumption of energy. Costs of these environmental damages generally are not incorporated into prices for energy commodities and resources; this omission leads to overuse of energy. Concern about this issue is common to energy economics, environmental economics, and ecological economics, with energy economics and environmental economics literature attempting to assign monetary valuation of the impacts and ecological economics rejecting the idea that a monetary value could be placed on environmental impacts.

Environmental impacts currently receiving most attention are associated with the release of greenhouse gases into the atmosphere, primarily carbon dioxide, from combustion of fossil fuels. The three primary fossil fuels – coal, petroleum, and natural gas – each include carbon. During combustion, carbon combines with oxygen to produce carbon dioxide, the primary greenhouse gas. Carbon dioxide accumulates in the atmosphere and is expected to result in significant detrimental impacts on the world's climate, including global warming, rises in the ocean levels, increased intensity of tropical storms, and losses in biodiversity.

Energy use leads to additional environmental damages. Coal combustion, particularly high sulfur coal combustion, emits oxides of sulfur, which, through atmospheric chemical reactions, result in acid rain. Automobile gasoline combustion releases oxides of nitrogen and volatile organic compounds, which, in the presence of sunlight, result in smog. Electric generating facilities often use much water for cooling and release the heated water into lakes or oceans, leading to local impacts on the ecosystem. Extraction of oil or mining of coal can lead to subsidence of the land overlying of the extracted deposits.

Pervasive environmental impacts of energy use, absent governmental intervention, imply that significant costs of energy use are not included in the price energy users face. These so-called externalities lead to overuse of energy and provide strong motivation for interventions designed to reduce energy use.

Energy Efficiency

Energy efficiency means cutting down the amount of energy required to perform an action like switching on a light, opening a fridge or heating water. For example, every time you flick on a light switch, you're using energy to light the room

But if you change from using an old style lightbulb, known as an incandescent lightbulb, to a more efficient fluorescent bulb - you'll be producing the same amount of light, but will be using less energy in the process. That's energy efficiency in a nutshell.

There's now a huge variety of products, services and designs that can help improve energy efficiency. Running older appliances, like washing machines, tumble dryers or fridge-freezers can account for a huge percentage of your energy bill, so it makes sense to ensure these are as economical as possible.

However, it's the little adjustments that are often the easiest for reducing your energy usage. These small changes may not seem like much, but every little does really help when it comes to saving energy.

Importance of Energy Efficiency

There are three main reasons to be more energy efficient:

The Environment

If you're making better use of the energy you're using, then less energy will need to be created. As a result, the power plants that make this energy will burn fewer natural fossil fuels, these limited natural resources will last longer and we're less likely to impact the earth by sourcing them through mining, logging or extractions.

Also, by reducing energy usage, the power plants will create fewer emissions of greenhouse gasses such as sulphur dioxide or carbon dioxide - which absorb the sun's heat and trap it in our atmosphere. So, less energy and less fuel means less damage to the natural world.

Cost Saving

This is a no-brainer, the less energy you use to light and heat your home, the lower the cost of your energy bills. Any steps you make will have an impact, although choosing energy efficient products and appliances is a great way to cut down your usage without having to make drastic changes that effect your daily routines.

The Global Economy

Dwindling natural resources can have an impact on the global economy. As oil, coal and natural gasses are depleting, they're becoming more expensive, meaning more strain on the societies that rely on them. Becoming more efficient means we waste fewer of these precious resources. Plus, as we become more efficient in our energy use at home, we're also more likely to be efficient across other kinds of industries, for example, businesses, agencies and schools. This means these areas can spend less on energy and more on goods, services, education and products.

Benefits

Improve Indoor Comfort and Keep Those Toes Warm!

A model energy efficiency building controls the flow of air, heat and moisture. Insulation, ENERGY STAR certified windows, and air sealing work to keep the air inside the building warm and dry. Mechanical ventilation systems (fans, heat recovery ventilators and air exchangers on heating systems) circulate the internal air as well as expel unwanted moisture and other pollutants. Properly sized, high-efficiency furnaces or boilers use less energy to heat buildings and, by controlling the air flow throughout the building, minimize the occurrences of hot and cold spots across all rooms.

Save Money on Energy Bills

Improving your home or building's energy efficiency will save you money. For example, the average homeowner participating in the Live Smart BC grant program will reduce energy use by approximately 30 percent, providing financial savings of $600 on a $2000 annual heating bill.

Breathe Healthy, Fresh, Clean Air

Just think of all the kettles, fish tanks, plants and even people who are expelling moisture into the air. Plus, microwaves, photocopiers, printers and even carpet and paint fumes all pollute indoor areas where we live and work. Controlling the airflow with mechanical ventilation helps confirm that the air in the building is being refreshed with outside air as often as required by building codes. For example, the air in a home should be completely replaced with fresh air every three hours. Fresh air takes less energy to heat than the inside air that slowly becomes contaminated with moisture and odors. An energy efficient and well-ventilated home is a healthy home.

Reduce Moisture Issues

10 to 50 litres of moisture are released inside a house each day. Humidity levels over 40% can create air quality problems. City Green advisors will show you how to reduce condensation and fogging on windows, staining on walls and ceilings, peeling paint, mould growth and odors.

When humidity is above 50%, airborne diseases become more difficult to control. To protect health and houses, all moist, stale air should be replaced with fresh air every 3 to 4 hours.

Finding the balance between minimizing air leaks and ensuring appropriate ventilation is critical. City Green's Certified Energy Advisors are trained to assess your home's energy needs based on a 'house as a system' building science approach. Only the "house as a system" building science approach to retrofit planning will prevent problems that single product installations may cause.

Increase Resale Value

Energy efficient buildings typically have longer lifecycles, lower maintenance fees, and cost less to operate. Rental properties are more comfortable to live in, are quickly occupied, and report lower tenant turnover rates. The natural light, comfortable temperatures and increased indoor air quality of energy efficient commercial buildings results in increased staff productivity. And, homeowners benefit from increased control over their energy bills while living in a home that supports their lifestyle needs.

Reduce Greenhouse Gas Emissions and Contribute to a Healthier Planet!

Through energy efficiency retrofits, the average home built 20 years ago has the potential to lower their energy bills by 30% – and to reduce the amount of greenhouse gasses the house produces by 4.5 tons per year. Lowering the total amount of greenhouse gasses in the atmosphere is an important step to reducing the impacts of climate change, both locally and globally.

Energy Efficiency vs Energy Conservation

Whilst energy efficiency and energy conservation both have a very similar goal, they are two different ways of approaching the saving of electricity and gas. Read on to find out more about the differences between them and their benefits.

Energy Efficiency

- Tends to apply to physical upgrades, rather than major lifestyle changes
- Uses technology which requires less energy to perform a certain function
- Can be more costly initially, but requires less effort in the long term
- Examples include: replacing old style lightbulbs with CFLs, adding insulation, upgrading appliances

Energy Conservation

- Involves a change in behaviour to use less energy
- Has the potential to save more energy, although requires a greater commitment
- Means saving energy without having to pay for new appliances or services
- Examples include: turning off electronics overnight, turning down the thermostat, leaving lights off longer.

Energy Efficiency Gap

Energy efficiency leads to a slower diffusion of energy-efficient products than would be expected if consumers made all positive net present value investments. This phenomenon has come to be known as the energy efficiency gap or the energy efficiency paradox. In some cases, the energy efficiency gap is defined even more broadly to describe the slower than socially optimal rate of diffusion of energy-efficient products. Yet the very existence of the energy efficiency gap has been met with skepticism by many economists and the use of analyses showing positive net present value energy efficiency investments for policy support has caused even more skepticism. The underlying issues for determining the existence of an energy efficiency gap are whether individual decision making is modeled correctly and whether all relevant costs are accounted for.

Reason for Small Energy Efficiency Gap

Despite more than thirty years of research on the energy efficiency gap, the issue of its size remains unresolved. Many economists believe that consumer choices reveal more about the economics of energy efficiency improvements than do engineering calculations. If engineering estimates of the energy savings potential from seemingly cost-effective investments fail to include some costs or model the consumer's decision inappropriately, then the assessment of what is optimal from the consumer's perspective will be incorrect. Thus the engineering approach will result in the net benefits from energy efficiency investments being overstated, which means the gap may be much smaller than estimated or there may be no gap at all. More broadly, the gap may be overestimated because of hidden costs, consumer heterogeneity, uncertainty, overestimated savings, and the rebound effect.

Hidden Costs

Hidden costs may be as simple as the administrative costs of an energy efficiency program or the time costs of finding or installing a more energy-efficient product that is as reliable as the more familiar alternative product. Hidden costs also include the opportunity cost of the services or alternative investments that consumers forgo in order to invest in energy efficiency. One potentially important opportunity cost is a decrease in the quality of the energy service provided. For example, more energy-efficient lighting may come at the cost of less pleasing or lower quality light. Higher vehicle fuel economy may be bundled with less desirable attributes such as small size or less acceleration. The evaluation of energy efficiency investments may also cause firms to divert scarce managerial attention from projects that may be more important to the firm.

Barriers for Energy Efficiency Gap

Energy efficiency gaps exist because market failures exist. It is important to identify and understand those barriers in order to achieve desirable government policy interventions. According to Hirst and Brown, various barriers that prevent the society from successfully closing energy efficiency gap can be divided into two categories: structural barriers and behavioral barriers. Structural barriers result from the actions of public and private organizations, and are usually beyond the control of the individual energy end user. Some examples are presented as following:

Distortion in fuel prices: The fuel prices that consumers pay do not reflect the social and environmental costs associated with fuel production, distribution and consumption. Consumers tend not to invest on energy efficiency technologies due to this distortion.

Uncertainty about future fuel prices: There have been great uncertainties with the prices for fuels, such as electricity and petroleum. More stringent environmental regulations and global warming concerns also increase the volatility of fuel prices. These uncertaintics prevent consumers from making rational purchase decisions of new energy-using systems.

Limited access to capital: Consumers often face high up-front costs for energy-efficient systems. In addition, high discount rates are used to make tradeoffs between the initial capital investment and reduced operating costs, which also hinder the investments in energy-efficiency technologies.

Government fiscal and regulatory policies: Government policies tend to encourage energy consumption, rather than energy efficiency. For instance, government support has focused more on energy production, and the profit of electric utilities is a function of sales.

Codes and standards: The development of codes and standards often lag behind the development of technologies. It also takes a long time to adopt and modify standards, which becomes a barrier for energy efficiency technological innovation.

Supply infrastructure limitations: The deployment of energy efficiency technologies is highly restricted by factors such as geography, infrastructure and human resources.

Behavioral barriers are problems that characterize the end-user's decision-marking relating to energy consumption. Four examples are discussed below.

Attitudes toward energy efficiency. Public's awareness of and attitudes toward energy efficiency could greatly affect their energy-related purchase and consumption behaviors.

Perceived risk of energy-efficiency investments. Consumers and businesses can be very risk-averse in terms of investing in energy efficiency technologies. The uncertainties of fuel prices and high discount rate for operating costs have both made energy-efficiency investments even more "risky" for many decision makers.

Information gaps. There is often a lack of information on the performance of energy-efficient technologies. Consumers tend not to change their energy consumption behavior if little information is provided.

Misplaced incentives. The principal-agent problem and a lack of life-cycle thinking on costs and savings have imposed barriers for energy conservation.

Jaffe and Stavins categorize the barriers differently. They think that both market failures and non-market failures could account for the limited market success of the cost-effective energy-efficiency technologies. One important source of market failure is imperfect information, for instance the public good attributes of information and information asymmetry. Non-market failure may include the heterogeneity and inertia of consumers, and uncertainty about future energy prices and actual savings from energy efficiency investments.

Measures Narrowing the Energy Efficiency Gap

Energy efficiency gap exists in various sectors, ranging from households, small businesses, corporations, and governments. Many policies and programs have been developed to overcome those barriers and close the energy efficiency gap. Subsidies and incentives for energy-efficient technologies. Insufficient capital investment could be overcome by more aggressive tax subsidies, loan guarantees, and low-interest government loans for energy efficient technologies.

Minimum building and equipment efficiency standards. Minimum building and equipment efficiency standards are cost-effective approaches to save energy. Effective implementation and upgrading of building energy efficiency standards could improve the energy integrity of new buildings, while equipment efficiency standards could help reduce energy consumption and pollution during the life-cycle of equipment.

Information programs. Research has proved that providing accurate and trustworthy information on energy use and energy efficiency choices could help narrow the gap. Three forms of information programs can be implemented to help producers and consumers make more informed and rational decisions. The first one is generic information applicable to all energy decision, such as forecasts of future energy prices; the second type of program is to provide comparative information to facilitate technology and product choices, such as product rating and labeling systems; the third type of program is to offer specific recommendations for producers' and consumers' investment choices or behavior changes.

Government procurement programs for energy-efficient technologies. Government agencies could be required to procure energy-efficient products. This would help improve the energy efficiency of government sector, and the "learning by doing" impact would create early markets for energy-efficient technologies. Some real-world examples of those measures include the following: EU's energy consumption labeling scheme, U.S. DOE's building energy codes program, and EPA's and DOE's ENERGY STAR® voluntary labeling programs.

Fuel and Fuel Efficiency

Any substance that can be consumed to produce energy is a fuel. This energy that releases is generally in the form of chemical energy or heat energy. The recent invention of nuclear technology means now even nuclear energy may be released due to nuclear fission or fusion.

This heat energy that fuels release is used for various purposes such as cooking, in heaters, for many industrial and manufacturing purposes. At other times we use an engine to convert this heat energy to mechanical energy. Like when we use petrol to run our cars.

And then there is the fuel our bodies use. Every cell requires energy to perform its functions. They get this energy from organic molecules such as carbohydrates, fats etc. This process of using fuels is known as cellular respiration. And these organic molecules are obtained via nutrition, which is why we call food as the fuel of our bodies.

Types of Fuel

Let us now look at some type of fuels:

- Solid Fuel: These are solid materials that combust to produce energy. Some examples of Solid fuel are coal, charcoal, soot, wood etc. These were most likely the first fuels utilized by mankind. They were the fuels responsible for the invention of fire. Even today they have very widespread household and industrial uses. Charcoal is still a very important fuel source for all manufacturing products and power industry, And wood is still widely used in houses to cook food and provide warmth
- Liquid Fuel: These are the fuels we burn to produce mechanical energy and kinetic energy. Most liquid fuels such as crude oil form due to exposure to intense heat and pressure to fossilized remains of plants and animals. Then there are biofuels in liquid form such as ethanol and hydrogen fuel. These fuels are easy to transport and relatively easy to use,
- Fuel Gas: Fuel Gas as the name suggests are fuels that are in a gaseous state under normal conditions. Some examples are methane, carbon monoxide, propane etc. They have an advantage that they can be easily transported to the place of consumption. However, they also tend to leak from pipes and every precaution must be taken to avoid this. The best example would be the CNG gas that comes to your kitchen via pipes that you utilize for cooking
- Biofuel: Biofuel can be solid, liquid or a gas. The only condition is that it must be derived

from Biomass. So essentially it is fuel derived from living matter that can be replenished. One such example is ethanol made from sugarcane wastes

- Fossil Fuel: These are the conventional fuels. They are also non-replenishable. They have formed over thousands of years in the earth's core where fossilized remains of animals and plants have been exposed to high pressure and temperatures. These are fast depleting and our dependence on them is a major concern for all of humanity.

Fuel Efficiency

Now you know that energy cannot be created or destroyed, All energy can only be transferred. Fuels also do not create energy. They only convert the chemical energy of the fuel to the kinetic energy with the help of the thermal energy supplied to them. The efficiency with which the fuel does this conversion of energy is known as fuel efficiency.

Fuel efficiency is measured as the amount of heat that 1 kg of fuel (any fuel) produces on combustion. This is known as the calorific value of the fuel. The unit of measurement of fuel efficiency is kilojoules per kg, i.e. kJ/kg.

Question: What is a Calorific Value

a. The Amount of heat energy produced on complete combustion of 1 Kg of fuel

b. Amount of heat energy produced on complete combustion of 100 Kg of fuel

c. The Amount of heat energy produced on complete combustion of 1 g of fuel

d. Amount of heat energy lost on complete combustion of 1 Kg of fuel

Solution: The correct option is "A". Calorific value is the quantity of heat produced by the complete combustion of a given mass of a fuel, usually expressed in joules per kilogram. It can be defined as the amount of heat energy produced on complete combustion of 1 Kg of fuel.

Characteristics of an Ideal Fuel

The characteristics of good fuel are as follows

- High calorific value.

- Proper ignition temperature (neither be too low nor too high)
- Cause no pollution to the atmosphere on combustion.
- A controllable rate of combustion.
- Easily available in plenty and cheap in cost.
- Easy to handle and transport.

Benefits to the Environment

Cleaner Air

A car that does not consume a lot of gas and gives the best possible mileage is a fuel-efficient car. How does a fuel-efficient car give cleaner air? The lesser the fuel consumption of the car, lesser is the smoke that will come out of its exhaust. And this is because the tailpipe emissions coming from fuel-efficient cars are lesser than the cars that are not fuel efficient. As a result, the damage done to the atmosphere is considerably less.

Gases, like carbon dioxide, carbon monoxide, nitrogen oxide, hydrocarbons, and particulate matter are let out by all cars. These gases are harmful to us as well as our environment. They aid in depleting our ozone layer, causing global warming, and bringing about climatic changes. Fuel-efficient cars help in reducing pollution and smog by at least 50%.

Reduces Dependency on Fossil Fuels

By switching over to a hybrid car or one that does not require too much fuel, we are reducing our dependency on fossil fuels. Fuel-efficient cars need less fuel to run, so the amount of gas that we use up in commuting reduces drastically.

In some countries, alternatives to non-renewable fuel, like propane and biodiesel in automobiles are used. Oil exploration will reduce, specially in environment sensitive areas, like the Arctic National Wildlife Refuge, thus helping to save our environment and wildlife living in those areas.

Imbibing Environmental Awareness

By choosing to drive a fuel-efficient car, you are making yourself as well as others around you aware of the benefits that these cars have on the environment. Invariably, you are spreading the word around that our environment needs care and that you are doing something about it.

Many car showrooms today are drawing light towards our environmental issues and promoting the use of cars that are fuel efficient. Discussing factors, like fuel economy, environmental protection, and energy conservation–which are some of the key selling points of these vehicles–help in increasing environmental awareness.

When we are planning to buy a car, we should always ask the dealer the exact engine specifications, like the size of the engine, number of cylinders used, weight and size of the vehicle, etc., as all these factors affect the fuel efficiency of the car. Cars with smaller engines burn lesser fuel, but some big-sized cars, such as those made of lightweight material like high-strength steel, fiber glass, aluminum, plastic, and other composite material are fuel efficient. When you buy a hybrid car, or even a car that is known to be fuel efficient, you know that you are doing the environment and yourself a favor. By conserving our environment, we are securing our future.

Energy Efficiency in Transport

In the modern world, commuting has become a big point of our daily lives as we live increasingly far distances from places of work - just a look at local highways during the morning and afternoon commute will show traffic that looks like the congestion illustrated in Fig. 1. As of 2013, the average commute to work is approximately 25.8 minutes, which is almost an additional five minutes to that of the 1980s. This additional amount of travel time may not seem too daunting, but time (and energy used) adds up quickly. Similarly, the most common mode of transportation of commuters is driving alone. For the purposes of this analysis, we will look at some of the most common types of automated forms of transportation.

Fig. An image of traffic congestion that could be alleviated with more efficient transportation.

This will include transportation by private automobile, urban transit bus, commuter rail, or domestic air travel. We will be examining the efficiency of each method of transportation by referring to their passenger miles per diesel equivalent gallon (DEG), which also takes into account how many people, on average, will be transported via this method and factors it into its energy use efficiency. Thus, the larger the number listed below in each category, the more fuel efficient it is considered.

This was calculated by first converting the total annual fuel used by each mode to units of DEG:

$$\text{Annual DEG} = \frac{\text{Fuel Energy Content (btu/gal)}}{\text{Diesel Energy Contgent (btu/gal)}} \times \text{Annual Fuel (gal)}$$

Thus, giving of the DEG to calculate the energy intensity measurement mentioned above (Pass-mi/DEG):

Passenger Miles per DEG (pass-mi/DEG)	=	$\frac{\text{Annual Passenger Miles}}{\text{Annual DEG}}$

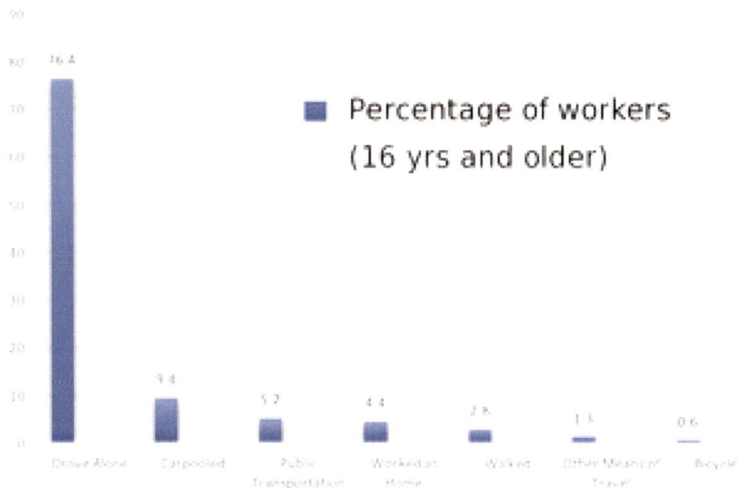

Fig. Percentage of workers using each mode of transportation. (Source: S. Tran)

With this in mind, the efficiency of the highlighted methods of transportation have been broken down and calculated as follows:

- Private automobile: As mentioned above, this is the largest sector for commuting in America. This category has a high pass-mi/gal of 84.1 and a low of 25.2, resulting in an average of 27.9 pass-mi/gal when considering only lone drivers. When carpooling (two people or more) is taken into account this average increases to an average of 55.9 pass-mi/gal (high of 121.1 and low 36.3).
- Urban Transit Bus: This mode of transportation only comprises 2.52% of the commuters , and includes the use of primarily diesel- powered, rubber-tired vehicles for fixed route scheduled service within an urban area. It has a high of 250 pass-mi/gal and a low of 3.6 pass-mi/gal, with an average of 70.5 pass-mi/gal.
- Commuter Rail: Transportation that uses electric or diesel-powered locomotives that pull passenger cars for local, short- distance travel between a central city and adjacent suburbs. It has a high of 169.5 pass-mi/gal and a low of 32.1 pass-mi/gal, with an average of 90.3 pass-mi/gal.
- Domestic Air Travel: This includes flights and it has an average of 54.8 pass- mi/gal.

Land Transport Means

Walking

A 68 kg (150 lb) person walking at 4 km/h (2.5 mph) requires approximately 210 kilocalories (880 kJ) of food energy per hour, which is equivalent to 4.55 km/MJ. 1 US gal (3.8 L) of gasoline contains about 114,000 British thermal units (120 MJ) of energy, so this is approximately equivalent to 360 miles per US gallon (0.65 L/100 km).

Nordic walkers

Velomobile

Velomobiles have the highest energy efficiency of any known mode of personal transport. At a speed of 50 km/h (31 mph), the velomobile manufacturer WAW claims that only 0.5 kW·h (1.8 MJ) of food energy per 100 km is needed to transport the passenger (= 18 J/m). This is around $\frac{1}{5}$ (20%) of what is needed to power a standard upright bicycle without aerodynamic cladding, and $\frac{1}{50}$ (2%) of that which is consumed by an average fossil fuel or electric car (the velomobile efficiency corresponds to 4700 miles per US gallon, 2000 km/L, or 0.05 L/100 km). Other sources give a figure of $\frac{1}{3.4}$ (29.5%) of the energy use of a normal bicycle.

Bicycle

A Chinese Flying Pigeon bicycle

A standard lightweight, moderate-speed bicycle is one of the most energy-efficient forms of transport. Compared with walking, a 64 kg (140 lb) cyclist riding at 16 km/h (10 mph) requires about half the food energy per unit distance: 27 kcal/km, 3.1 kW·h (11 MJ) per 100 km, or 43 kcal/mi. This converts to about 732 mpg_{US} (0.321 L/100 km; 879 mpg_{imp}). This means that a bicycle will use between 10-25 times less energy per distance traveled than a personal car, depending on fuel source and size of the car. This figure does depend on the speed and mass of the rider: greater speeds give higher air drag and heavier riders consume more energy per unit distance. In addition, because bicycles are very lightweight (usually between 7–15 kg) this means they consume very low amounts of materials and energy to manufacture. In comparison to an automobile weighing 1500 kg or more, a bicycle typically requires 100-200 times less energy to produce than an automobile.

Motorized Bicycle

A motorized bicycle allows human power and the assistance of a 49 cm^3 (3.0 cu in) engine, giving a range of 160 to 200 mpg_{US} (1.5–1.2 L/100 km; 190–240 mpg_{imp}). Electric pedal-assisted bikes run on as little as 1.0 kW·h (3.6 MJ) per 100 km, while maintaining speeds in excess of 30 km/h (19 mph). These best-case figures rely on a human doing 70% of the work, with around 3.6 MJ (1.0 kW·h) per 100 km coming from the motor. This makes an electric bicycle one of the most efficient possible motorized vehicles, behind only a motorized velomobile.

Electric kick Scooter

Electric kick scooters, part of a scooter-sharing system, in San Jose, California.

Electric kick scooters, such as those used by scooter-sharing systems like Bird or Lime, typically have a maximum range of under 30 km (19 mi) and a maximum speed of roughly 15.5 mph (24.9 km/h). Intended to fit into a last mile niche and be ridden in bike lanes, they require little skill from the rider. Because of their light weight and small motors, they're extremely energy-efficient.

Human Power

To be thorough, a comparison must also consider the energy costs of producing, transporting and packaging of fuel (food or fossil fuel), the energy incurred in disposing of exhaust waste, and the energy costs of manufacturing the vehicle. This last can be significant given that walking requires little or no special equipment, while automobiles, for example, require a great deal of energy to produce and have relatively short lifespans. In addition, any comparison of electric vehicles and liquid-fuelled vehicles must include the fuel consumed in the power station to generate the electricity. In the UK for instance the efficiency of the electricity generation and distribution system is around 0.40.

Automobiles

The automobile is an inefficient vehicle compared to other modes of transport. This is because the ratio between the mass of the vehicle and the mass of the passengers is much higher when compared to other modes of transport.

Bugatti Veyron

Automobile fuel efficiency is most commonly expressed in terms of the volume of fuel consumed per one hundred kilometres (L/100 km), but in some countries (including the United States, the United Kingdom and India) it is more commonly expressed in terms of the distance per volume fuel consumed (km/L or miles per gallon). This is complicated by the different energy content of fuels such as petrol and diesel. The Oak Ridge National Laboratory (ORNL) states that the energy content of unleaded gasoline is 115,000 British thermal unit (BTU) per US gallon (32 MJ/L) compared to 130,500 BTU per US gallon (36.4 MJ/L) for diesel.

A second important consideration is the energy costs of producing energy. Bio-fuels, electricity and hydrogen, for instance, have significant energy inputs in their production. Hydrogen production efficiency are 50-70% when produced from natural gas, and 10-15% from electricity. The efficiency of hydrogen production, as well as the energy required to store and transport hydrogen, must to be combined with the vehicle efficiency to yield net efficiency. Because of this, hydrogen automobiles are one of the least efficient means of passenger transport, generally around 50 times as much energy must be put into the production of hydrogen compared to how much is used to move the car.

A third consideration to take into account when calculating energy efficiency of automobiles is the occupancy rate of the vehicle. Although the consumption per unit distance per vehicle increases with increasing number of passengers, this increase is slight compared to the reduction in consumption per unit distance per passenger. This means that higher occupancy yields higher energy efficiency per passenger. Automobile occupancy varies across regions. For example, the estimated average occupancy rate is about 1.3 passengers per car in the San Francisco Bay Area, while the 2006 UK estimated average is 1.58.

Fourth, the energy needed to build and maintain roads is an important consideration, as is the energy returned on energy invested (EROEI). Between these two factors, roughly 20% must be added to the energy of the fuel consumed, to accurately account for the total energy used.

Finally, vehicle energy efficiency calculations would be misleading without factoring the energy cost of producing the vehicle itself. This initial energy cost can of course be depreciated over the life of the vehicle to calculate an average energy efficiency over its effective life span. In other words, vehicles that take a lot of energy to produce and are used for relatively short periods will require a great deal more energy over their effective lifespan than those that do not, and are therefore much less energy efficient than they may otherwise seem. Hybrid and electric cars use less energy in their operation than comparable petroleum-fueled cars but more energy is used to manufacture them, so the overall difference would be less than immediately apparent. Compare, for example, walking, which requires no special equipment at all, and an automobile, produced in and shipped from another country, and made from parts manufactured around the world from raw materials and minerals mined and processed elsewhere again, and used for a limited number of years. According to the French energy and environment agency ADEME, an average motor car has an embodied energy content of 20,800 kWh and an average electric vehicle amounts to 34,700 kWh. The electric car requires nearly twice as much energy to produce, primarily due to the large amount of mining and purification necessary for the rare earth metals and other materials used in lithium-ion batteries and in the electric drive motors. This represents a significant portion of the energy used over the life of the car (in some cases nearly as much as energy that is used through the fuel that is consumed, effectively doubling the car's per-distance energy consumption), and cannot be ignored when comparing automobiles to other transport modes. It is important to note, also, that as these are average numbers for French automobiles and they are likely to be significantly larger in more auto-centric countries like the United States and Canada, where much larger and heavier cars are more common.

Driving practices and vehicles can be modified to improve their energy efficiency by about 15%.

On a percentage basis, if there is one occupant in an automobile, between 0.4-0.6% of the total energy used is used to move the person in the car, while 99.4-99.6% (about 165 to 250 times more) is used to move the car.

Example Consumption Figures

Two American solar cars in Canada

- Solar cars use no externally supplied fuel other than sunlight, charging the batteries entirely from built-in solar panels, and typically use less than 3 kW·h per 100 miles (67 kJ/km or 1.86 kW·h/100 km). These cars are not designed for passenger or utility use and would not be practical as such due to speed, payload, and inherent design.
- The four passenger GEM NER uses 169 Wh/mi (203 mpge; 10.5 kW·h/100 km), which equates to 2.6 kW·h/100 km per person when fully occupied, albeit at only 24 mph (39 km/h).
- The General Motors EV1 was rated in a test with a charging efficiency of 373 Wh-AC/mile or 23 kWh/100 km approximately equivalent to 2.6 L/100 km (110 mpg_{imp}; 90 mpg_{US}) for petroleum-fuelled vehicles.

- Chevrolet Volt in full electric mode uses 36 kilowatt-hours per 100 miles (810 kJ/km; 96 mpge), meaning it may approach or exceed the energy efficiency of walking if the car is fully occupied with 4 or more passengers, although the relative emissions produced may not follow the same trends if analyzing environmental impacts.
- The Daihatsu Charade 993cc turbo diesel (1987–1993) won the most fuel efficient vehicle award for going round the United Kingdom consuming an average of 2.82 L/100 km (100 mpg_{imp}). It was surpassed only recently by the VW Lupo 3 L which consumes about 2.77 L/100 km (102 mpg_{imp}). Both cars are rare to find on the popular market. The Daihatsu had major problems with rust and structural safety which contributes to its rarity and the quite short production run.
- The Volkswagen Polo 1.4 TDI Bluemotion and the SEAT Ibiza 1.4 TDI Ecomotion, both rated at 3.8 L/100 km (74 mpg_{imp}; 62 mpg_{US}) (combined) were the most fuel efficient petroleum-fueled cars on sale in the UK as of 22 March 2008.[needs update]
- Honda Insight – achieves 48 mpg_{US} (4.9 L/100 km; 58 mpg_{imp}) under real-world conditions.
- Honda Civic Hybrid- regularly averages around 45 mpg_{US} (5.2 L/100 km; 54 mpg_{imp}).
- 2012 Cadillac CTS-V Wagon 6.2 L Supercharged, 14 mpg_{US} (17 L/100 km; 17 mpg_{imp}).
- 2012 Bugatti Veyron, 10 mpg_{US} (24 L/100 km; 12 mpg_{imp}).
- 2018 Honda Civic: 36 mpg_{US} (6.5 L/100 km; 43 mpg_{imp})
- 2017 Mitsubishi Mirage: 39 mpg_{US} (6.0 L/100 km; 47 mpg_{imp})
- 2017 Hyundai Ioniq hybrid: 55 mpg_{US} (4.3 L/100 km; 66 mpg_{imp})
- 2017 Toyota Prius: 56 mpg_{US} (4.2 L/100 km; 67 mpg_{imp}) (Eco trim)
- 2018 Nissan Leaf: 30 kWh (110 MJ)/100 mi (671 kJ/km) or 112 MPGe
- 2017 Hyundai Ioniq EV: 25 kWh (90 MJ)/100 mi (560 kJ/km) or 136 MPGe

Trains

Trains are in general one of the most efficient means of transport for freight and passengers. Efficiency varies significantly with passenger loads, and losses incurred in electricity generation and supply (for electrified systems), and, importantly, end-to-end delivery, where stations are not the originating final destinations of a journey.

Actual consumption depends on gradients, maximum speeds, and loading and stopping patterns. Data produced for the European MEET project (Methodologies for Estimating Air Pollutant Emissions) illustrate the different consumption patterns over several track sections. The results show the consumption for a German ICE high-speed train varied from around 19 to 33 kW·h/km (68–119 MJ/km; 31–53 kW·h/mi). The data also reflects the weight of the train per passenger. For example, TGV double-deck Duplex trains use lightweight materials, which keep axle loads down and reduce damage to track and also save energy.

The specific energy consumption of the trains worldwide amounts to about 150 kJ/pkm and 150 kJ/tkm (i.e. 4,2 kWh/100 pkm and 4,2 kWh/100 tkm) in terms of final energy. Passenger transportation by rail requires less than one-tenth of the energy needed to move a person by car or plane. This is the reason why, although accounting for 9% of world passenger transportation activity (expressed in pkm) in 2015, rail passenger services represented only 1% of final energy demand in passenger transportation.

Freight

Energy consumption estimates for rail freight vary widely, and many are provided by interested parties. Some are tabulated below.

Country	Year	Fuel economy (weight of goods)	Energy Intensity
USA	2007	185.363 km/L (1 short ton)	energy/mass-distance
UK	—	87 t·km/L	0.41 MJ/t·km (LHV)

Passenger

Country	Year	Train efficiency	Per passenger-km (kJ)	Note
China	2018	9.7 MJ (2.7 kWh) /car-km	137 kJ/passenger-km (at 100% load)	CR400AF@350km/h Beijing-Shanghai PDL 1302km average
Japan	2004	17.9 MJ (5.0 kWh)/car-km	350 kJ/passenger-km	JR East average
Japan	2017	1.49 kWh/car-km	~92 kJ/passenger-km	JR East Conventional Rail
EC	1997	18 kW·h/km (65 MJ/km)		
USA		1.125 mpg_{US} (209.1 L/100 km; 1.351 mpg_{imp})	468 passenger-miles/US gallon (0.503 L/100 passenger-km)	
Switzerland	2011	2300 GWhr/yr	470 kJ/passenger-km	
Basel, Switzerland		1.53 kWh/vehicle-km (5.51 MJ/vehicle-km)	85 kJ/passenger-km (150 kJ/passenger-km at 80% average load)	
USA	2009		2,435 BTU/mi (1.60 MJ/km)	
Portugal	2011	8.5 kW·h/km (31 MJ/km; 13.7 kW·h/mi)	77 kJ/passenger-km	

Considering only the energy spent to move the train, and taking as example the urban area of Lisbon, train seems to be on average 7 times more efficient than automobile for transport of passengers, if we consider energy spent per passenger-km. Considering an automobile which has a consumptions of around 6 l/100 km (47 mpg_{imp}; 39 mpg_{US}) of gasoline, the fact the on average cars in Europe have an occupation ratio of around 1.2 passengers per automobile and that one litre of gasoline amounts for about 8826 Wh, one gets on average 441 Wh (1,590 kJ) per passenger-km. On the other hand, a modern urban train with an average occupation of 20% of total capacity, which has a consumption of about 8.5 kW·h/km (31 MJ/km; 13.7 kW·h/mi), requires 21,5 Wh of energy per passenger-km, 1/20th of that of the automobile.

Braking Losses

Stopping is a considerable source of inefficiency. Modern electric trains like the *Shinkansen* (the *Bullet Train*) use regenerative braking to return current into the catenary while they brake. A Siemens study indicated that regenerative braking might recover 41.6% of the total energy consumed. The Passenger Rail (Urban and Intercity) and Scheduled Intercity and All Charter Bus Industries Technological and Operational Improvements – FINAL REPORT states that "Commuter operations can dissipate more than half of their total traction energy in braking for stops." and that "We estimate head-end power to be 35 percent (but it could possibly be as high as 45 percent) of total energy consumed by commuter railways." Having to accelerate and decelerate a heavy train load of people at every stop is inefficient despite regenerative braking which can recover typically around 20% of the energy wasted in braking. Weight is a determinant of braking losses.

Buses

The Bus Rapid Transit of Metz uses a diesel-electric hybrid driving system, developed by Belgian Van Hool manufacturer.

- In July 2005, the average occupancy for buses in the UK was stated to be 9.
- The fleet of 244 40-foot (12 m) 1982 New Flyer trolley buses in local service with BC Transit in Vancouver, Canada, in 1994/95 used 35,454,170 kWh for 12,966,285 vehicle km, or 9.84 MJ/vehicle km. Exact ridership on trolleybuses is not known, but with all 34 seats filled this equates to 0.32 MJ/passenger km. It is quite common to see people standing on Vancouver trolleybuses. This is a service with many stops per kilometre; part of the reason for the efficiency is the use of regenerative braking.
- A commuter service in Santa Barbara, California, USA, found average diesel bus efficiency of 6.0 mpg_{US} (39 L/100 km; 7.2 mpg_{imp}) (using MCI 102DL3 buses). With all 55 seats filled this equates to 330 passenger mpg; with 70% filled, 231 passenger mpg.
- In 2011 the fleet of 752 buses in the city of Lisbon had an average speed of 14.4 km/h and an average occupancy of 20.1 passengers per vehicle.

Other

- NASA's Crawler-Transporter is used to move the Shuttle from storage to the launch pad. It uses diesel and has one of the highest fuel consumption rates on record, 150 US gallons per mile (350 l/km; 120 imp gal/mi).

Air Transport Means

Aircraft

Solar Impulse 2, a solar aircraft

A principal determinant of energy consumption in aircraft is drag, which must be opposed by thrust for the aircraft to progress.

- Drag is proportional to the lift required for flight, which is equal to the weight of the aircraft. As induced drag increases with weight, mass reduction, with improvements in engine efficiency and reductions in aerodynamic drag, has been a principal source of efficiency gains in aircraft, with a rule-of-thumb being that a 1% weight reduction corresponds to around a 0.75% reduction in fuel consumption.
- Flight altitude affects engine efficiency. Jet-engine efficiency increases at altitude up to the tropopause, the temperature minimum of the atmosphere; at lower temperatures, the Carnot efficiency is higher. Jet engine efficiency is also increased at high speeds, but above about Mach 0.85 the airframe aerodynamic losses increase faster.
- Compressibility effects: beginning at transonic speeds of around Mach 0.85, shockwaves form increasing drag.
- For supersonic flight, it is difficult to achieve a lift to drag ratio greater than 5, and fuel consumption is increased in proportion.

Concorde fuel efficiency comparison (assuming jets are filled to capacity)		
Aircraft	Concorde	Boeing 747-400
Passenger-miles/imperial gallon	17	109
Passenger-miles/US gallon	14	91
Litres/100 passenger-km	16.6	3.1

Passenger airplanes averaged 4.8 l/100 km per passenger (1.4 MJ/passenger-km) (49 passenger-miles per gallon) in 1998. Note that on average 20% of seats are left unoccupied. Jet aircraft efficiencies are improving: Between 1960 and 2000 there was a 55% overall fuel efficiency gain (if one were to exclude the inefficient and limited fleet of the DH Comet 4 and to consider the Boeing 707 as the base case). Most of the improvements in efficiency were gained in the first decade when jet craft first came into widespread commercial use. Compared to advanced piston engine airliners of the 1950s, current jet airliners are only marginally more efficient per passenger-mile. Between 1971 and 1998 the fleet-average annual improvement per available seat-kilometre was estimated at 2.4%. Concorde the supersonic transport managed about 17 passenger-miles to the Imperial gallon; similar to a business jet, but much worse than a subsonic turbofan aircraft. Airbus puts the fuel rate consumption of their A380 at less than 3 l/100 km per passenger (78 passenger-miles per US gallon).

Air France Airbus A380-800

The mass of an aircraft can be reduced by using light-weight materials such as titanium, carbon fiber and other composite plastics. Expensive materials may be used, if the reduction of mass justifies the price of materials through improved fuel efficiency. The improvements achieved in fuel efficiency by mass reduction, reduces the amount of fuel that needs to be carried. This further reduces the mass of the aircraft and therefore enables further gains in fuel efficiency. For example, the Airbus A380 design includes multiple light-weight materials.

Airbus has showcased wingtip devices (sharklets or winglets) that can achieve 3.5 percent reduction in fuel consumption. There are wingtip devices on the Airbus A380. Further developed Minix winglets have been said to offer 6 percent reduction in fuel consumption. Winglets at the tip of an aircraft wing smooth out the wing-tip vortex (reducing the aircraft's wing drag) and can be retrofitted to any airplane.

NASA and Boeing are conducting tests on a 500 lb (230 kg) "blended wing" aircraft. This design allows for greater fuel efficiency since the whole craft produces lift, not just the wings. The blended wing body (BWB) concept offers advantages in structural, aerodynamic and operating efficiencies over today's more conventional fuselage-and-wing designs. These features translate into greater range, fuel economy, reliability and life cycle savings, as well as lower manufacturing costs. NASA has created a cruise efficient STOL (CESTOL) concept.

Fraunhofer Institute for Manufacturing Engineering and Applied Materials Research (IFAM) have researched a shark skin imitating paint that would reduce drag through a riblet effect. Aircraft are a major potential application for new technologies such as aluminium metal foam and nanotechnology such as the shark skin imitating paint.

Propfan propulsors are a more fuel efficient technology than jets or turboprops, but turboprops have an optimum speed below about 450 mph (700 km/h). This speed is less than used with jets by major airlines today. However, the decrease in speed reduces drag. With the current *[needs update]* high price for jet fuel and the emphasis on engine/airframe efficiency to reduce emissions, there is renewed interest in the propfan concept for jetliners that might come into service beyond the Boeing 787 and Airbus A350XWB. For instance, Airbus has patented aircraft designs with twin rear-mounted counter-rotating propfans. NASA has conducted an Advanced Turboprop Project (ATP), where they researched a variable pitch propfan that produced less noise and achieved high speeds.

Related to fuel efficiency is the impact of aviation emissions on climate.

Small Aircraft

Dyn'Aéro MCR4S

- Motor-gliders made of composite materials can reach an extremely low fuel consumption for cross-country flights, using thermal air currents and winds.
- At 160 km/h, a diesel powered two-seater Dieselis burns 6 liters of fuel per hour, 1.9 liters per 100 passenger km.
- at 220 km/h, a four-seater 100 hp MCR-4S burns 20 liters of gas per hour, 2.2 liters per 100 passenger km.
- Under continuous motorised flight at 225 km/h, a Pipistrel Sinus burns 11 liters of fuel per flight hour. Carrying 2 people aboard, it operates at 2.4 liters per 100 passenger km.
- Ultralight aircraft Tecnam P92 Echo Classic at cruise speed of 185 km/h burns 17 liters of fuel per flight hour, 4.6 liters per 100 passenger km (2 people). Other modern ultralight aircraft have increased efficiency; Tecnam P2002 Sierra RG at cruise speed of 237 km/h burns 17 liters of fuel per flight hour, 3.6 liters per 100 passenger km (2 people).
- Two-seater and four-seater flying at 250 km/h with old generation engines can burn 25 to 40 liters per flight hour, 3 to 5 liters per 100 passenger km.

- The Sikorsky S-76C++ twin turbine helicopter gets about 1.65 mpg_{US} (143 L/100 km; 1.98 mpg_{imp}) at 140 knots (260 km/h; 160 mph) and carries 12 for about 19.8 passenger-miles per gallon (11.9 L per 100 passenger km).

Water Transport Means

Ships

Queen Elizabeth

Queen Elizabeth

Cunard stated that Queen Elizabeth 2 travelled 49.5 feet per imperial gallon of diesel oil (3.32 m/l or 41.2 ft/US gal), and that it had a passenger capacity of 1777. Thus carrying 1777 passengers we can calculate an efficiency of 16.7 passenger miles per imperial gallon (16.9 l/100 p·km or 13.9 p·mpg_{-US}).

Cruise Ships

MS *Oasis of the Seas* has a capacity of 6,296 passengers and a fuel efficiency of 14.4 passenger miles per US gallon. Voyager-class cruise ships have a capacity of 3,114 passengers and a fuel efficiency of 12.8 passenger miles per US gallon.

Emma Maersk

Emma Maersk uses a Wärtsilä-Sulzer RTA96-C, which consumes 163 g/kW·h and 13,000 kg/h. If it carries 13,000 containers then 1 kg fuel transports one container for one hour over a distance of 45 km. The ship takes 18 days from Tanjung (Singapore) to Rotterdam (Netherlands), 11 from Tanjung to Suez, and 7 from Suez to Rotterdam, which is roughly 430 hours, and has 80 MW, +30 MW. 18 days at a mean speed of 25 knots (46 km/h) gives a total distance of 10,800 nautical miles (20,000 km).

Assuming the Emma Maersk consumes diesel (as opposed to fuel oil which would be the more precise fuel) then 1kg diesel = 0.832 litres = 0.317 US gallons. This corresponds to 46,525 kJ. Assuming a standard 14 tonnes per container (per teu) this yields 74 kJ per tonne-km at a speed of 45 kmh (24 knots).

Sailboats

A sailboat, much like a solar car, can locomote without consuming any fuel. A sail boat such as a Dinghy using just wind power requires no input energy in terms of fuel. However some manual energy is required by the crew to steer the boat and adjust the sails using lines. In addition energy will be needed for demands other than propulsion, such as cooking, heating or lighting. The fuel efficiency of a single-occupancy boat is highly dependent on the size of its engine, the speed at which it travels, and its displacement. Due to the high viscosity of water, with a single passenger, the equivalent energy efficiency will be lower than in a car, train, or plane.

Energy Saving Tips in Transportation

- Don't use your private vehicles extensively. Try to use public transportation whenever possible.
- Remove extra BOOT weight in the car. Unwanted items like a flat tyre, unwanted luggage in the vehicle, especially heavy ones, can reduce the vehicle's mileage. The reduction is based on the percentage of extra weight (relative to the vehicle's weight) and affects smaller vehicles more than the bigger ones.
- Keep automobiles fuel filters clean and save the fuel.
- Don't forget to keep your vehicle tuned up. When a vehicle is running well, it uses nine per cent less fuel and thus emits fewer toxic and noxious fumes.
- Don't use clutch pedal as footrest.
- Beyond one minute, it is more fuel - efficient to restart your car.
- Maintain recommended tyre pressure.

References

- Combustion and-fuel/introduction-to-fuel-and-fuel-efficiency, chemistry: toppr.com, Retrieved 28 May 2018
- Benefits-energy-efficiency: citygreen.ca, Retrieved 19 June 2018
- Environmental-benefits-of-fuel-efficient-cars-that-all-should-know-13954: matteroftrust.org, Retrieved 10 March 2018
- Energy-efficiency, help-and-support: npower.com, Retrieved 14 March 2018
- Energy-conservation-in-transport-systems, energy-conservation-in-transport-sector: vikaspedia.in, Retrieved 31 April 2018

Permissions

All chapters in this book are published with permission under the Creative Commons Attribution Share Alike License or equivalent. Every chapter published in this book has been scrutinized by our experts. Their significance has been extensively debated. The topics covered herein carry significant information for a comprehensive understanding. They may even be implemented as practical applications or may be referred to as a beginning point for further studies.

We would like to thank the editorial team for lending their expertise to make the book truly unique. They have played a crucial role in the development of this book. Without their invaluable contributions this book wouldn't have been possible. They have made vital efforts to compile up to date information on the varied aspects of this subject to make this book a valuable addition to the collection of many professionals and students.

This book was conceptualized with the vision of imparting up-to-date and integrated information in this field. To ensure the same, a matchless editorial board was set up. Every individual on the board went through rigorous rounds of assessment to prove their worth. After which they invested a large part of their time researching and compiling the most relevant data for our readers.

The editorial board has been involved in producing this book since its inception. They have spent rigorous hours researching and exploring the diverse topics which have resulted in the successful publishing of this book. They have passed on their knowledge of decades through this book. To expedite this challenging task, the publisher supported the team at every step. A small team of assistant editors was also appointed to further simplify the editing procedure and attain best results for the readers.

Apart from the editorial board, the designing team has also invested a significant amount of their time in understanding the subject and creating the most relevant covers. They scrutinized every image to scout for the most suitable representation of the subject and create an appropriate cover for the book.

The publishing team has been an ardent support to the editorial, designing and production team. Their endless efforts to recruit the best for this project, has resulted in the accomplishment of this book. They are a veteran in the field of academics and their pool of knowledge is as vast as their experience in printing. Their expertise and guidance has proved useful at every step. Their uncompromising quality standards have made this book an exceptional effort. Their encouragement from time to time has been an inspiration for everyone.

The publisher and the editorial board hope that this book will prove to be a valuable piece of knowledge for students, practitioners and scholars across the globe.

Index

A

Acid Rain, 8, 12, 26, 40, 226, 235

Air Pollution, 8-9, 12, 18, 26, 29, 36, 45, 54, 80

Artificial Photosynthesis, 203, 210

B

Battery Energy Storage System, 86, 115

Bell-type Gasholder, 117-118

Binary Cycle, 39-40, 199, 214, 216, 218-220, 224

Biodiesel, 4, 67, 205, 207-208, 243

Biofuel, 204-208, 241

Biological Methanation, 131, 133

Biomass Energy, 4, 25, 39, 50, 53-54, 204-206, 211, 213

Biomethane, 133, 208

Building Life Cycle, 157, 159-161

C

Centralized Gas Turbine, 32, 34

D

Domestic Energy Consumption, 55, 71

E

Electric Power Consumption, 82, 84

Electricity Generation, 13-14, 16, 18, 20, 27-28, 45, 51, 56, 84, 90, 162, 198-200, 202, 205, 208-209, 212, 216-217, 222, 224, 227, 234, 247, 250

Energy Commodities, 228, 230-232, 234

Energy Conservation Technique, 168

Energy Consumption, 3-4, 14, 55, 63-72, 75-80, 92, 127, 147-149, 152, 154-159, 162, 165, 167, 177-179, 197-198, 206, 215, 239-241, 249, 251, 253

Energy Conversion, 19, 45, 59, 69-70, 81, 101, 128, 131, 140, 228-231, 233-234

Energy Data, 84, 149

Energy Economics, 194, 228-229, 233-234

Energy Efficiency, 31, 34-35, 53, 62, 69, 73-74, 76-77, 79-81, 84, 112, 154, 156, 163-164, 169, 177-178, 191, 199, 228, 231, 235-241, 244, 246, 248-250, 257

Energy Efficiency Gap, 228, 238-240

Energy Management Challenge, 147, 154, 165

Energy Management System, 65, 157, 165-166

Energy Policy, 46, 59-64, 84-85, 147, 198

Energy Procurement, 153-154

Energy Recycling, 190, 192

Energy Resources, 3, 59-60, 62-64, 84, 87, 154, 167, 169, 178, 199, 215, 224, 228-230, 232-233

Energy Strategy, 63, 147, 155-156

F

Facility Management, 147, 151-152, 154, 157-158, 161, 163

Flash Steam Power Plant, 219, 226

Flywheel Energy Storage, 86, 88, 110-111

Frequency Regulation, 94, 99, 114

Fuel Catalyst, 184-185

Fuel Cell, 69, 130-131, 135-145

Fuel Cell Efficiency, 139-140

Fuel Efficiency, 113, 128, 176, 184, 191-192, 228, 230-231, 241-243, 248, 253-257

Fuel Ionizer, 183

Fuel Saving Device, 167, 179, 184

G

Gas Holder, 118

Geothermal Energy, 2-3, 6, 39, 196, 199-200, 213-218, 221, 224-225, 228

Geothermal Power Plant, 40, 214, 217

Geothermal Resource, 199, 214

Geothermal System, 6, 222

Greenhouse Gas Emission, 49

Grid Electricity, 89, 93, 212

H

Heat Rate, 167, 170-172, 174-175

Heat Rate Deviation, 172

Hydroelectric Dam, 93, 202

Hydroelectricity, 7, 14, 46, 88, 94, 145, 202-203, 213

Hydropower, 1, 3, 11, 46, 48-49, 97, 100-101, 130, 195-198, 201-203, 209, 211, 213

I

Industrial Energy Management, 147, 163

Intermittency, 86-87, 210

N

National Energy Policy, 60-61, 64

Non Renewable Energy, 1, 15

Nuclear Power, 1-2, 10-11, 16-20, 25-26, 36, 38, 46, 64, 94, 145, 156, 196-197

P

Payback Period, 73-74, 150, 155

Piston Gasholder, 120-121

Plant Heat Rate, 170-172

Power Generation, 17-20, 33-34, 126, 190-191, 209, 212

Pumped-storage, 86, 88, 94, 96, 100-102, 146

Pyrolysis, 51-52

R

Radiator Reflector, 193

Reactor Core, 17-18, 36-37

Rebound Effect, 167, 175-179, 239

Renewable Resource, 10, 12-13, 25, 44, 196, 200, 214, 224

S

Smes System, 102-104, 106-107

Solar Energy, 1-2, 4, 41, 43, 45, 67, 72, 90, 93, 131, 195, 198, 203-204, 215, 217, 224, 226-227, 231

Superconducting Magnetic Energy Storage, 89, 102

Superinsulation, 167, 188-189

T

Telescopic Gasholder, 118-119

Thermal Energy Storage, 93-94, 122-126, 191, 217

Thermal Power, 16, 93, 126, 191, 198, 215

Thermal Storage, 90, 94-95, 189, 191

Thermodynamic Efficiency, 69-70, 185

Tidal Energy, 6-7, 63, 86, 208-209

Total Primary Energy Supply, 58-59, 78

W

Water Pollution, 4, 26, 29, 133

Wind Power, 4, 12, 16, 39, 43-46, 90, 100, 130, 133, 140, 144, 195, 198, 200-201, 208, 210-211, 257

Wind Turbine, 14, 43-44, 46, 90, 211

World Energy Consumption, 55, 67-68